TIGER TANK

OSPREY
PUBLISHING

TIGER TANK

MARCUS COWPER

First published in Great Britain in 2016 by Osprey Publishing,
PO Box 883, Oxford, OX1 9PL, UK
PO Box 3985, New York, NY 10185-3985, USA
E-mail: info@ospreypublishing.com

OSPREY PUBLISHING, PART OF BLOOMSBURY PUBLISHING PLC

In the compilation of this volume we relied on the following Osprey titles:
DUE 2, *Sherman Firefly vs Tiger* by Stephen A. Hart; DUE 37, *King Tiger vs IS-2* by David R. Higgins; NVG 1, *King Tiger Heavy Tank 1942–1945*; and NVG 5, *Tiger I Heavy Tank 1942–45* by Tom Jentz and Hilary Doyle.

A CIP catalogue record for this book is available from the British Library

ISBN: 978 1 4728 1294 0
ePub ISBN: 978 1 4728 1296 4
PDF ISBN: 978 1 4728 1295 7

Index by Zoe Ross
Maps on page 82 and 95 by Peter Bull Art Studio, and on page 126 by Bounford.com
Typeset in Sabon and Cleanwork
Originated by PDQ Media, Bungay, UK
Printed in China through World Print Ltd.

16 17 18 19 20 10 9 8 7 6 5 4 3 2 1

ILLUSTRATIONS

Artworks on pages 22, 25, 32, 35, 70–71, 114 and 129 by Peter Sarson (© Osprey Publishing); pages 41, 49, 53, 54, 55, 91, 99, 119 and 124 by Jim Laurier (© Osprey Publishing).

Front cover: The Tiger I driven by Tiger ace Michael Wittmann when engaged by a Sherman Firefly in August, 1944. (Jim Laurier, © Osprey Publishing)

Title page: A formation of Tiger II tanks, end of 1944. (Ullstein Picture via Getty Images)

Contents page: View of a Tiger I, early 1940s. This one seems to be somewhat damaged as it is missing track guards as well as one of its front fenders. The image was originally published as '*Das Heer im Grossdeutschen Freiheitskampf*' (translated as 'The Army in the Greater German Freedom Struggle'), a collection of 50 plus images taken by the German Army's combat photography unit (*Propagandakompanie*). (Photo by Schroter/PhotoQuest/Getty Images)

Back cover: Tiger II with series production turret. (Jim Laurier, © Osprey Publishing)

Osprey Publishing supports the Woodland Trust, the UK's leading woodland conservation charity. Between 2014 and 2018 our donations will be spent on their Centenary Woods project in the UK.

www.ospreypublishing.com

The following will help in converting measurements:	
1km = 0.62 miles	1cm = 0.39 inches
1kg = 2.25lb	1 metric tonne = 1.1 US (short)
1m = 0.91 yards	tons = 0.98 UK (long) tons
	1l = 0.26 US gal = 0.22 Imp gal

CONTENTS

INTRODUCTION

The Tiger I and II are arguably the most famous tanks of World War II, if not the most famous tanks in the entire history of armoured warfare.

The Tiger I was, perhaps surprisingly for a tank which achieved such a reputation, quickly designed utilising components that had been partially tested in previous heavy Panzers. The need for a new and better-armoured heavy tank that would be successful in combating British tanks and anti-tank guns had become increasingly evident to the Germans in the first years of the war. Underlining the need for an improved tank, the Soviet T-34 medium tank, encountered after the Nazi invasion of Russia on 22 June 1941, shocked the Germans with its thick, sloped skin, excellent mobility, and powerful armament. Following the appearance of

the T-34 and the heavy Soviet KV-1, the design and production of an effective heavy Panzer was pursued with increased urgency. By the time the first models rolled off the production line in August 1942, the Tiger I featured extremely thick armour, providing it with what was at that time a formidable level of battlefield survivability. It also mounted a powerful long-barrelled 8.8cm gun that could at normal combat ranges defeat virtually every enemy tank then in existence. The launch of Tiger II production in January 1944 allowed the deployment of the most powerful combat tank during the war. Its thick, sloped armour made it virtually impervious to any Allied tank or gun.

Germany's Tigers dominated the battlefields of Europe between late 1942 and early 1945, striking fear

OPPOSITE Tigers on the Eastern Front, between Bielgorod and Orel, July 1943. (Photo by Roger Viollet/Getty Images)

into those Allied crews unfortunate enough to encounter them on the battlefield; many such crews did not survive these invariably brief and bloody actions. Although relatively few in quantity, the numbers of Tigers available allowed German forces to slow the rising tide of Allied battlefield success for longer than they would have otherwise been able to.

As a result of the rapid arms race in the East, where each side attempted to maintain a battlefield edge, vehicle weight, armour protection and firepower all increased. In April 1944 the new Soviet IS-2 tank entered the field; its 122mm round packed considerable force, and in combat it proved well suited for its task as a breakthrough tank that could stand up to the German Tigers. The British, too, had been developing a tank to counter the Tiger threat: the Sherman Firefly, mounting a potent 17-pounder gun which made it an effective opponent for even the heavily armoured Tigers at normal combat ranges. By summer 1944, the new German Tiger II had been deployed, and would soon be sent to the Eastern Front. The Tiger II was more often known by its unofficial name, '*Königstiger*' ('Bengal Tiger'), incorrectly translated by Allied intelligence as 'King Tiger' or 'Royal Tiger'. By expanding on the thick armour and large main armament of the Tiger I, and the more modern design of the Panther, the 70-tonne Tiger II presented a formidable battlefield solution. Despite the advances in Allied tank design, Tiger I and II tanks would remain deadly battlefield opponents to Allied armoured units right up until the closing days of the war.

OPPOSITE A Tiger II in fighting position in the forest, 1945. (Ullstein Picture via Getty Images)

CHRONOLOGY

1937
January   Henschel is contracted to develop a heavy breakthrough tank that eventually becomes the Tiger I.

1940
Autumn   Dr Ferdinand Porsche is commissioned to develop a 45-tonne tank.

1941
26 May   Meeting with Hitler at which development of both Henschel and Porsche designs are accelerated and requirements are established for armour, speed and gun calibre and penetration. Krupp is tasked with producing the armament and turrets for these models.

22 June   Operation *Barbarossa* begins – the German invasion of the Soviet Union.

1942
20 April   Field trials of the rival Henschel-Tiger VK45.01(H) and Porsche-Tiger VK45.01(P) prototypes are held at Rastenburg, East Prussia, in front of Hitler in honour of his 53rd birthday. The Henschel design triumphs.

July   The Henschel design is tested extensively at Bad Berka, Germany. Henschel is contracted to mass-produce the winning design under the designation Panzerkampfwagen VI Tiger Ausf. E.

OPPOSITE One of the first Tigers to enter the war, in combat between destroyed Soviet tanks south of Lake Ladoga near Leningrad in September 1942. (Ullstein Bild via Getty Images)

20 August   The first four main-production run Tigers are completed at Henschel's Kassel plant.

29 August   The Tiger I is first used in combat with 502nd Heavy Panzer Battalion near Leningrad.

1943

January   Hitler orders Porsche and Henschel to begin development work on a new Tiger tank, to be ready for production in February 1943.

2 February   Krupp's prototype (later 'series production') turret is delivered for testing at the Kummersdorf research facility. It will later be used on Tiger II models.

July–December   Porsche develops the VK45.02(P) and Henschel the VK45.03(H) models. Henschel's model triumphs again, and will become known as the Tiger II.

1944

January   Production of the Tiger II begins.

February   The first Tiger II tanks roll off the production line; hereafter Tiger IIs increasingly replace the Tiger I in German heavy tank units.

14 March   The first Tiger IIs are issued to Panzer Lehr Division's 316th Panzer Company (Fkl) as radio-control vehicles for the tracked Borgward BIV *Sprengstoffträger* ('explosive carrier').

April   Soviet IS-2 tanks are first deployed in combat as part of the 11th and 72nd Guards Heavy Tank regiments.

6 June   D-Day: Allied forces land on the coast of Normandy, in German-occupied France, and the Tiger encounters the Sherman Firefly.

18 July   Operation *Atlantic* in Normandy sees Tiger IIs first used in combat, as part of 1st Company, 503rd Heavy Panzer

Battalion.

13 August   Tiger IIs are first used on the Eastern Front, at the Sandomierz bridgehead, as the 501st Heavy Panzer Battalion clashes with IS-2s of the 71st Guards Heavy Tank Regiment.

Late August   The last Tiger I (out of a total production run of 1,349) is completed by Henschel.

1945

12 January   Tiger IIs from the 424th Heavy Panzer Battalion confront IS-2s of the 13th Guards Heavy Tank Regiment near Lisów, Poland. Operation *Konrad* sees IS-2s and Tiger IIs of Heavy Panzer Battalion Feldherrnhalle clash during the siege of Budapest.

29 March   The final Tiger II leaves the factory.

April–May   IS-2s and Tiger IIs (502nd and 503rd Heavy SS-Panzer battalions) fight against each other for the final time during the battle of Berlin.

Early May   The last Tiger tanks still operational – probably fewer than 30 – surrender as Germany's armed forces capitulate.

DESIGN AND DEVELOPMENT

THE TIGER I

The Tiger can trace its direct development back to 1941, and its indirect antecedents to 1937. During the period 1937–40, the Germans carried out development work on a tank heavier than their then heaviest tank – the Panzer IV. By 1940, this programme for a 30-tonne tank – designated the VK30.01 – had produced several prototype designs, named the Breakthrough tanks (*Durchbruchwagen*) 1 and 2, or DW1 and DW2, and the VK30.01(H).

The DW1 chassis, developed by the German armaments firm Henschel and Son, sported 50mm-thick armour plates and was powered by a 280bhp Maybach ML 120 engine. Its suspension featured the typical German torsion bar suspension. During 1939 Henschel produced its DW2 design. This tank married a modified DW1 chassis to a Krupp-designed turret that mounted the 7.5cm KwK 37 L/24 gun used in the Panzer IV. Finally, during 1940 Henschel produced the VK30.01(H) design. This comprised a turretless chassis which incorporated a novel running-gear arrangement based on interleaved wheels. The 30-tonne VK30.01(H), which featured 600mm-thick armour plates, was powered by a 300bhp Maybach HL 116 six-cylinder engine and could reach a maximum road speed of 34kph.

During 1940, the Henschel firm also began work on the heavier VK65.01(H) design – an enlarged, up-armoured and larger-engined version of the Panzer IV designed to fulfil the army's future 65-tonne Panzer

OPPOSITE Henschel's VK30.01(H) design, seen here in early 1940. This design departed from traditional German design by the introduction of a novel running-gear arrangement based on interleaved road wheels. This successful arrangement subsequently led Henschel to use this type of arrangement in their Tiger prototype, the VK45.01(H). (Tank Museum)

requirement. This massive design was to be powered by a 12-cylinder 600bhp Maybach HL 224 engine. Developmental work on this design did not progress any further, however, because the *Waffenamt* (WaA – the Army Weapons Agency) was happy with the Panzer IV as its heaviest vehicle. Nevertheless, these design efforts influenced the subsequent work that would lead to the production of the Tiger.

The rival armaments firm of Porsche, meanwhile, had begun to develop a heavy tank designated the VK45.01(P) – seemingly at Hitler's request and without formal contracts from the Army Weapons Agency. From spring 1941, Krupp which had been instructed to collaborate with both Henschel and Porsche on the heavy-tank project, supplied Porsche with its recently developed 8.8cm KwK 36 L/56 tank gun – a modified version of the famous 88mm (3.46in.) anti-aircraft gun. The long-barrelled Krupp tank gun delivered an impressive anti-tank capability by achieving a high muzzle velocity for its rounds.

The efforts to produce an effective heavy tank design received fresh impetus on 26 May 1941, when a meeting of experts chaired by Hitler reviewed future German tank development strategy. At this meeting, the Führer demanded that a well-armoured German heavy tank be developed that would outgun any enemy tank it might encounter. Future development work should proceed on the basis of a vehicle that sported 100mm-thick frontal armour and a gun that could penetrate 100mm of armour at a range of 1,500m. The Führer thus ordered that the work already undertaken by the firms of Porsche and Henschel should be accelerated so that each could construct six prototype vehicles by the summer of 1942.

During the second half of 1941, the German Panzers were shocked by their first encounter with two unexpectedly formidable Soviet AFV designs – the T-34

OPPOSITE The interleaved running wheels before assembly. These provided stability and a smooth ride, but made maintenance difficult as all the outer wheels had to be taken off to access the inner ones and the torsion bars. (Panzerfoto)

medium and KV heavy tanks. These modern Soviet tanks outclassed all German tanks then in existence, including their heaviest vehicle, the Panzer IV. This realisation added new urgency to the development of new German medium and heavy tank designs.

Hitler wanted the future heavy tank to mount an 8.8cm gun – either the KwK 36 or a version of the new and yet more powerful Rheinmetall-Borsig 8.8cm FlaK 41 L/74. After experimentation, Porsche concluded that the latter weapon was not suitable for mounting in the turret it was then developing. The Army Weapons Agency, on the other hand, felt that mounting such a large gun in a tank (which would need to be large to accommodate a turret with a sufficiently wide turret ring to house the gun) would render the vehicle too heavy and immobile. The agency felt that the future heavy tank should mount a smaller 60mm or 70mm tapered-bore gun. This was a gun with an interior bore to the barrel that narrowed towards the muzzle. This narrowing squeezed the special tungsten-carbide round into the rifling on the inside of the barrel, enabling the round to be fired with greater muzzle velocity and accuracy, but crucially from a smaller gun. This was called the Gerlich principle after the engineer who had patented the idea in 1903. Tungsten, however, was already scarce and in much demand within the German war economy. The wrangling associated with the dispute over the choice of gun led to the simultaneous commencement of work on two separate prototype heavy tank programmes.

HENSCHEL VS PORSCHE

The Germans contracted one project, designated VK36.01(H), to Henschel based on a specification for a vehicle that mounted the tapered-bore 60mm or 70mm gun. Meanwhile, Porsche finally received formal contracts to produce the heavier VK45.01(P) design, which mounted Krupp's 8.8cm tank gun. By mid-1941 Henschel had produced seven prototype VK36.01(H) tanks that featured thick frontal armour and interleaved road wheels. However, the Germans then concluded that, such was the demand for and scarcity of tungsten, the tapered-bore gun central to the Henschel project was no longer feasible. The only way that the smaller Henschel heavy tank could compete with the heavier rival Porsche one in tank-killing capability was to employ tapered-bore technology; with this ruled out, Henschel had no choice but to abandon the VK36.01(H) programme. The Germans, however, did not wish to waste the valuable design work Henschel had put into this project. Consequently, the Army Weapons Agency contracted Henschel to develop a 41-tonne heavy tank, the VK45.01(H). Henschel decided to develop an enlarged version of its VK36.01(H) that would mount the same 8.8cm tank gun featured in the rival VK45.01(P) design. By late 1941, therefore, both Henschel and Porsche were now working on rival 41-tonne heavy tank designs that mounted the same gun.

By April 1942, Porsche and Henschel had completed their first prototype VK45.01(P) and VK45.01(H) heavy tanks, now generally referred to as the Porsche-Tiger and Henschel-Tiger, respectively. These rival designs had certain common features; most notably they mounted the same Krupp-designed turret that featured the 8.8cm KwK 36 L/56 gun and the co-axial MG 34 machine gun. Both designs also had a ball-mounted bow machine gun and sported heavy armour up to a maximum thickness of 120mm on the turret mantlet. Moreover, both tanks weighed around 55 tonnes, markedly above the original specification, because the German hierarchy increased the project's required levels of armour during the design process. Beyond this commonality, however, the two rivals were quite different.

The VK45.01(P) was powered by two air-cooled Simmering-Graz-Pauker (SGP) 320bhp engines that drove the tank through a series of dynamos and electric motors; the vehicle's drive mechanism was thus of petro-electric type, a typical Porsche arrangement. The VK45.01(P) also featured a novel suspension that comprised six steel double road wheels suspended in pairs from longitudinal torsion bars. Because of its high fuel consumption, however, the tank could only achieve a disappointing operational range of just 50km. The tank also featured a low squat chassis with the angular Krupp turret located well forward, which resulted in the long 8.8cm gun overhanging the front of the vehicle to a conspicuous degree. This made the design very heavy at the front end, and consequently the tank was prone to become bogged down in soft terrain. Nevertheless, during April 1942, before this design had even been evaluated, Porsche received contracts for 90 tanks to be delivered during the period January–April 1943.

RIGHT A Maybach HL 230 P45 engine about to be lowered into the hull of a Tiger I at Henschel's Kassel plant. (Panzerfoto)

EARLY TIGER I, 1ST COMPANY, 502ND HEAVY PANZER BATTALION, NEAR LENINGRAD, JANUARY 1943

Four Tigers of the 1st Company, 502nd Heavy Panzer Battalion were sent to the front near Leningrad and were the first to see action when committed near Mga on 29 August 1942. The rest of the company arrived at the front on 25 September, bringing the total number up to nine Tigers along with nine Panzer IIIs and nine Panzer III Ausf. Ns. All of these Panzers were initially painted *Feldgrau* ('field grey', a grey-green).

This Tiger I, tactical number 100, belonging to the company commander, was the first Tiger captured intact by the Soviets in January 1943. A large reproduction of the 502nd emblem, a *Mammut* (mammoth), had been painted on the rear of the turret of this Tiger. The stowage bin on the turret side was made and fitted by the troops at the front.

The first nine Tigers issued to the 502nd had *Fahrgestell* (Fgst – chassis) numbers 250002 through to 250010. Produced in August and September 1942,

these Tigers never possessed features that are commonly expected on a Tiger I. These Tigers did not have the removable mudguards on the superstructure sides, a track replacement cable on the left superstructure side, the toolbox at the left rear, or the hinged mudguards at the front and rear. However, they were unique in having an early track design for the Kgs 63/725/120 *Geländeketten* where the right track was a mirror image of the left track.

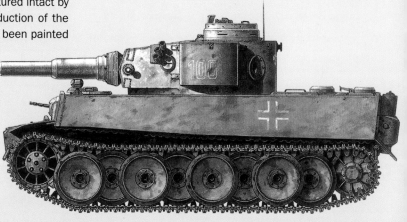

Several features differentiated Henschel's VK45.01(H) tank from its Porsche rival. The Henschel tank mounted the identical Krupp turret, but in the centre of the vehicle as opposed to the front. This arrangement reduced the degree to which the main gun overhung the front of the vehicle, rendering it less front-heavy. The Henschel tank's running gear featured the same novel interleaved road-wheel arrangement used in the earlier VK30.01(H). Unlike the Porsche design, the Henschel tank's suspension was based on the typical German arrangement of lateral torsion bars. Moreover, the tank's hull superstructure was wider and more angular than that of the Porsche tank. The VK45.01(H) also featured a single, rear-located 642bhp Maybach petrol engine, and its fuel tanks provided it with a marginally better cross-country operating range than its Porsche rival. The Henschel tank also featured a hydraulically controlled, preselected eight-speed Maybach Olvar gearbox and semi-automatic transmission.

To evaluate which of the two rival designs was superior, the Germans held a field trial in front of Hitler and other top Nazi officials at Rastenburg on 20 April 1942 – the Führer's 53rd birthday. When this and subsequent competitive trials had been completed, the army concluded – despite Hitler's prejudiced favouritism for Porsche – that the Henschel model was better than the Porsche; the greatest advantages were Henschel's superior engine power, reliability and vehicle mobility. The Army Weapons Agency also concluded that the Henschel tank was better suited for mass production than its rival – an important consideration for the already hard-pressed German war economy.

PRODUCTION

Hitler demanded that the new tank be committed to action as soon as possible. In July 1942 the Army Weapons Agency contracted Henschel to mass-produce this tank under the designation Panzer VI Ausf. E Tiger. Simultaneously, the agency cancelled the contracts already awarded to Porsche for its 90 Porsche-Tigers. Not wishing to waste these 90 partially constructed

TIGER I, 101ST HEAVY SS-PANZER BATTALION, NORMANDY, 1944

This Tiger I was produced after June 1944 and therefore features all of the final changes incorporated in the series. The fighting compartment was painted in *Elfenbein* (ivory) down to the level of the bottom of the sponson. Below this level no paint was applied over the red primer undercoat. Earlier vehicles in the series had this lower portion painted in olive green or *Feldgrau* (grey-green). The engine compartment was left in the red primer.

The external vertical surfaces of the Tiger were coated in *Zimmerit* anti-magnetic paste and the whole vehicle painted in dark yellow. The crew applied the camouflage stripes and patches of dark green and red brown. The tactical numbers of the 101st Heavy SS-Panzer Battalion (which was later renamed the 501st Heavy SS-Panzer Battalion) were painted in blue outlined in yellow. The unit insignia, crossed keys in a shield surrounded by oak leaves, was painted in white on the front and rear armour of most of the tanks in the unit.

This Tiger was issued to the 101st Heavy SS-Panzer Battalion and served in Normandy in the battles that followed the Allied landings and the subsequent breakout in the late summer of 1944.

The cutaway drawing shows the interior of the vehicle as it appeared in late production run variants. In particular it shows how the improved counterbalance spring was moved from its original position in the front right-hand side of the turret and relocated behind the gun and to the right of the commander's seat. This modification was introduced with the appearance of the new turret in July 1943. The more powerful Maybach HL 230 P45 engine shown replaced the Maybach HL 210 P45 from May 1943.

Changes were made to the stowage of items in the turret and electrical components were grouped on a panel on the right-hand side of the firewall between the fighting compartment and the engine.

KEY

1. *Zimmerit* anti-magnetic paste
2. Bosch headlight
3. 7.92mm MG 34 hull machine gun
4. Radio sets
5. Co-axial 7.92mm MG 34
6. Loader's periscope
7. Loader's hatch
8. Rear escape hatch
9. Main gun
10. Commander's cupola
11. Maybach HL 230 P45 12-cylinder 23-litre engine
12. Fuel tank (left side)
13. TzF 9c monocular sight
14. Steel road wheels with internal rubber cushioning
15. Drive sprocket
16. 725mm-wide battle tracks

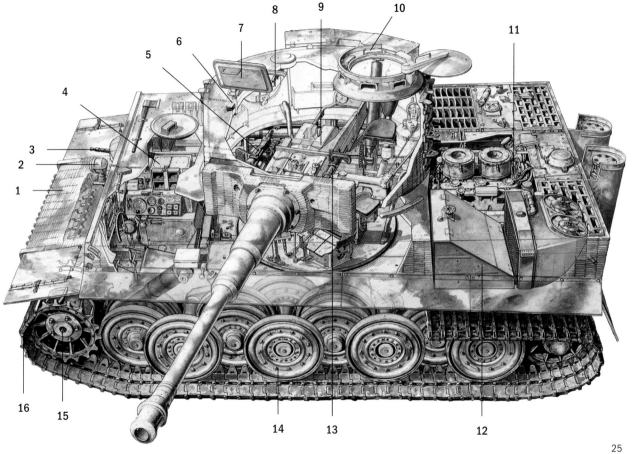

chassis, however, the Germans subsequently used them to produce an improvised heavy tank destroyer, the Panzerjäger Tiger (P) 'Elefant' (Elephant). In addition to its test vehicle (designated Experimental Series Panzer VI H1), the first two Tiger tanks Henschel produced in early August at Kassel were two pre-production vehicles. Henschel then completed the first four tanks in the main production run on 20 August. These were rushed off to the Eastern Front, and first entered combat on 29 August 1942 near Leningrad.

LAYOUT

The Tiger I employed a conventional internal layout for German tanks: in a forward open crew compartment, the driver sat on the left with the radio operator/machine gunner on the right, on either side of the gearbox. The central fighting compartment, containing the commander (at the back), the gunner (on the left of the gun) and the loader (on the right of the gun, facing backwards), was suspended from the turret by three steel tubes and revolved with the turret.

The engine was housed in a central compartment towards the rear of the tank, with the fans, radiator and fuel tanks in separate sections on either side of it.

VARIANTS

The Germans only developed two specialised Tiger I variants, which were both command-tank (*Befehlswagen*) designs. The only other significant Tiger I variant was the *Sturmtiger* assault vehicle, developed during 1943–44 to engage heavily fortified enemy bunkers.

Henschel produced 89 command-tank Tiger Is – either the SdKfz 267 battalion commander's tank or the SdKfz 268 company commander's vehicle. Both of these were similar to the standard production Tiger except for the addition of a powerful radio transmitter. The SdKfz 267 featured the ultra-long-range FuG 8

OPPOSITE A *Befehlstiger*, tactical number 003, featuring the distinctive Porsche hull and the additional aerial for command radio equipment. The turret is turned toward the rear of the tank. (Panzerfoto)

30-watt transmitter/medium-range receiver, while the SdKfz 268 mounted the long-range FuG 7 20-watt transmitter/ultra-short-wave receiver. Both command designs possessed a second aerial to service their additional communication sets, and this enabled friend and foe alike to distinguish these vehicles from standard Tiger Is. Space was created for these radios by reducing main gun-round stowage to 66 rounds and removing the co-axial machine gun.

THE TIGER II

ORIGINS

By mid-1942, heavy tank design had advanced considerably from when Henschel and Son began development of the 30-tonne *Durchbruchwagen* in 1937. While Germany had been on the strategic offensive between 1939 and 1942, lighter armoured vehicles such as the 23-tonne Panzer III and 25-tonne Panzer IV had proved sufficient in the manoeuvre and exploitation roles. Combat on the Eastern Front, however, necessitated tanks and self-propelled guns of increasing size and firepower as the Germans and Soviets both attempted to maintain an edge in the arms race. As Germany's military stance steadily transitioned to the defensive, the continued modification and modernisation of existing armoured vehicle types would not be a sustainable, long-term solution. An entirely new vehicle design was needed. This was to be the Panzer VI Ausf. B, or 'Tiger II' – also known to the Allies (due to a mistranslation) as the King Tiger.

Henschel's 57-tonne Tiger I had proved an effective counter to the Soviet T-34 and the heavier KV-1 and British designs, but by mid-war its boxy design was already ageing. The German authorities knew they would eventually need a more modern replacement, and Henschel and its rival Porsche were duly tasked with its development.

OPPOSITE These Tiger II tanks with Porsche turrets were issued to the 503rd Heavy Panzer Battalion in July 1944. In this photo they are preparing to train with live ammunition. (Panzerfoto)

On 26 May 1942, the Army Weapons Agency's *Waffenprüfamt* (Weapons Proving Office) 6 determined that the replacement for the Tiger I should be able to achieve 40kph, have a main armament capable of penetrating 100mm of rolled homogeneous armour (RHA) from 1,500m, and possess front and side armour of 150mm and 80mm respectively. Under Weapons Proving Office 6's head (Colonel Friedrich-Wilhelm Holzhäuer) and chief designer (Heinrich Ernst Kniepkamp), development soon got under way for what would become the heaviest operational tank of the war. Henschel expanded on their 45-tonne VK45.01(H), mounting the experimental 75mm/50mm tapering Waffe 0725; however, the scarcity of tungsten used in the projectile forced the cancellation of the weapon.

Porsche's VK45.02(P) proposal was based on its previous attempt to secure the Tiger I contract. However, the vehicle's weak engine and suspension, high ground pressure, and its over-engineered petrol/electrical drive train – which relied on copper and other materials that would be in short supply for wartime mass production

– sealed its fate. Intended to incorporate Rheinmetall-Borsig's new 8.8cm Flak 41 L/74 anti-aircraft gun, it was found to require a breech/counterweight that was too long to fit into turrets designed to house the shorter main armament of the Tiger I. Although Ferdinand Porsche held Hitler's favour, his efforts to produce the Tiger II were cancelled on November 3, 1942, in favour of the competition's improved prototype.

As part of this *Tigerprogram*, Henschel's updated VK45.03(H) possessed the long 8.8cm KwK 43 L/71 main gun, the most powerful tank gun available, coupled with sloped armour that resembled a bulked-up medium Mark V Panther. To simplify future maintenance and supply, the Tiger II also included transmission, track, engine cooling system, and other components that were interchangeable with the proposed Panther replacement. Hitler's calls for thicker armour and improved manoeuvrability meant that work on the prototype was slower than anticipated as the vehicle's side panels had to be strengthened during production to compensate for the added weight.

PRODUCTION

Since the 1800s Henschel had been a producer of locomotives, and the company was therefore in a position to build a host of combat-related armoured vehicles, trucks, aircraft and artillery in the lead-up to, and during, World War II. Its factory complex at Kassel comprised locomotive and gun production, a foundry and an armoured-vehicle assembly works where some 8,000 employees worked in two 12-hour shifts. Instead of an assembly line where skilled labourers assembled components in a sequential manner, Henschel organised production as a series of nine *Taktzeiten* (cycle times – *Takt* for short), where large vehicle sections were completed in a defined period before being moved along the line. Bombed some 40 times – often severely, as in a massive RAF raid on 22/23 October 1943 – the company continued to operate in at least a partial capacity until it was overrun by American ground forces on 4 April 1945.

In October 1943 planning was put in place to build 176 Tiger IIs (including three prototypes) at Henschel's factory between November 1943 and January 1944. In November, the run of the new tank was expanded by an additional 350, and the order was finally increased to 1,500 vehicles. As neither Henschel nor Porsche possessed the ability to construct the raw turrets and hulls, these were provided by Friedrich Krupp AG, which along with Dortmund Hörder Hutten Verein (DHHV) and Škoda Works of Czechoslovakia produced the main armour components. Turrets would then go to the Kassel-based Wegmann & Co. for final assembly before being sent to Henschel for mounting.

As Porsche's design requirement for a 1,900mm turret ring proved too small for the 8.8cm main armament, Krupp designed one with a 2,000mm diameter to provide a stable platform and accommodate both Porsche's and Henschel's lengthy main armaments with minimal modifications. Porsche had prematurely created 50 turrets for their VK45.02(P) prototype, and so the first Tiger II chassis were fitted with these. Externally the 'pre-production'

TIGER II, 'PRE-PRODUCTION' TURRET

The Replacement and Training Battalion 500 was equipped with some of the earliest production Tiger IIs. In this period all the Tigers were coated with *Zimmerit*. For camouflage the troops were issued with 2kg of dark green and red-brown paste which could be diluted with any petroleum based liquid or even water and applied with a spray in broad stripes and patches on top of the base dark yellow.

The regulations stated that the call-sign numbers were to be made up of numbers 30cm high, with black lines 3cm wide outlined in 1cm white. The first number is the *Kompanie* (company), one to three. For Tiger IIs in each company, the second digit is the *Zug* (platoon), one to three, and the third is the tank number, one to four. On the company commander's and deputy commander's Tigers the second digit was 0 and the third 0, 1 or 2. (There were 14 Tiger IIs per company.) The *Stabskompanie* (staff company) of each heavy tank battalion had three *Befehlswagen* Tiger IIs. These normally had the numbers 001 to 003 or in some units the Roman numerals I, II, and III were used. There was a total of 45 Tiger IIs per battalion (3+14+14+14).

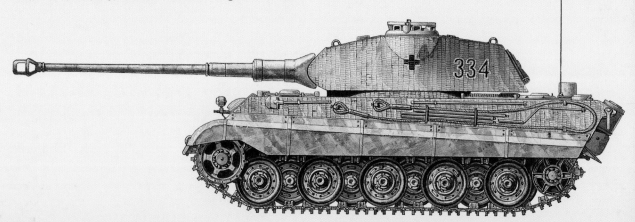

turret possessed a curved mantlet (similar to Panther D and A models), which proved resilient, but low impacts could damage the turret ring or deflect a round into the much thinner hull roof. This design shortcoming, and the rather involved construction process, ended the use of such 'pre-production' turrets on 7 December 1943. Krupp created the remaining 'series production' versions, which featured a *Saukopf* (pig's head) mantlet that minimised impact-related jamming of the main gun, and could accommodate many of Porsche's turret components with little or no alteration. Its more simplified armour configuration also resulted in greater internal space so that it held 86 main gun rounds instead of the 80 in Porsche's design.

LAYOUT

The Tiger II's interior and crew positions adhered to conventional German tank layout, with the forward compartment housing the driver and radio operator/bow machine gunner behind the glacis on the left and right, respectively. The driver had an adjustable seat, which could be raised to allow his head to protrude through his open hatch to improve visibility when driving in non-combat environments. Both crewmen had their own hatches and were separated by the vehicle's drive train and radio equipment.

The central fighting compartment was the turret, which, because of the vehicle's lengthy barrel, needed to be long and roomy to accept the breech's recoil and function as a counterweight. An attached platform/'basket' enabled the turret to rotate as a complete entity with its component parts and floor, which improved safety and operating efficiency. The gunner sat to the left, just ahead of the commander, with the loader to the breech's right. The turret's size limited the commander's downward visibility from the cupola when 'buttoned up' (with the hatches closed). As the Tiger II's transmission and drive wheels were at its front, a universal joint needed to be run under the turret basket to the rear-mounted engine, which increased the vehicle's height by 0.5m, and added to its weight as more armour was needed to cover the difference.

TIGER II, 'SERIES PRODUCTION' TURRET

This cutaway of a Tiger II shows the fighting compartment painted in *Elfenbein* (ivory). The motor compartment was left in the red primer undercoat. To simplify production, from Tiger II *Fahrgestell* (chassis) number 280177, the interior surfaces were not to be painted in *Elfenbein*. This meant that the red primer undercoat was the dominant colour. Certain other sub-assembly firms used *Feldgrau* (grey-green) on the items they manufactured. The fire extinguisher on the firewall between motor and fighting compartment was bright red.

The camouflage was applied at the factory and consisted of base red primer with overpainted stripes and patches of dark yellow and dark green (number 6003 in the RAL colour system) with sharp outlines. To simulate sunlight passing through foliage, all dark areas were painted with spots of dark yellow. Spots of green and red-brown were applied to the dark yellow areas, green nearest the green areas and red-brown near the red primer base.

KEY

1. Muzzle brake
2. 8.8cm KwK 43 L/71 gun
3. Driver's hatch
4. Gun mantlet
5. TzF 9b monocular sighting telescope
6. Loader's periscope
7. Loader's hatch
8. Ventilating fan
9. Racks of 8.8cm ammunition
10. Commander's cupola
11. Two-metre antenna for FuG 5 radio
12. Rear escape hatch
13. Pistol port
14. Commander's seat
15. Gunner's seat
16. Fuel tanks
17. Swing arm
18. Torsion bars
19. Driver's seat (could be raised for head-out driving)
20. Radio operator's position
21. 7.92mm MG 34
22. FuG 5 10-watt transmitter/ultra-short-wave receiver
23. Radio operator's periscope (fixed)
24. Driver's periscope (rotatable)

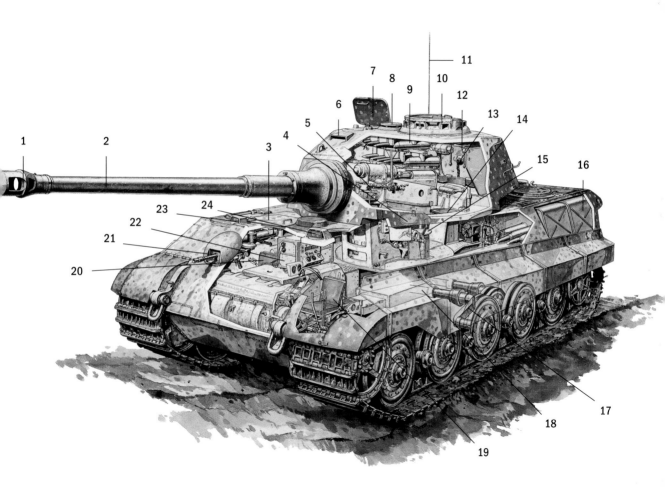

VARIANTS

Although the Tiger II had a 17-month production run, few significant changes were made to the basic design. Some vehicles in the first two series (420500 and 420530) had deep-fording submersion kits attached, but these were not used outside of testing. In January 1944 flat fenders were replaced with curved ones, straight exhaust pipes were bent to keep exhaust fumes from entering the engine and a device was added to heat the cooling water to improve starting in wintry conditions. In May, a new type of track was introduced that minimised uneven wear and helped prevent it from climbing over the drive sprocket when in motion. Factory-applied *Zimmerit* anti-magnetic paste was used on all four series runs until 9 September 1944; it was mistakenly believed to catch fire following projectile impacts, and was

discontinued. Because the vehicle's great weight placed considerable stress on gaskets, seals and sockets, such components were improved to prevent leaks and breakdowns.

Three *Pilz* ('mushroom') sockets were welded to the turret's roof to anchor the tripod of a two-tonne jib crane that was used to help move heavier vehicle components. In July, four track-link hangers were mounted on the turret's sides, and retrofitted to existing vehicles, and in the following month the 20-tonne jack was no longer issued. Starting in November 1944, Wegmann & Co. received 20 Tiger IIs for conversion into *Befehlswagen* (command vehicles) housing either an FuG 8 (SdKfz 267) or an FuG 7 radio (SdKfz 268) radio set, but these did not re-enter combat until February–March 1945. To prevent rain from obscuring the gunner's view, small shields were fitted over the sight's turret opening, starting in January 1945.

OPPOSITE A company of Tiger II tanks with Henschel turret and *Zimmerit* anti-magnetic paste, lined up in readiness for a propaganda film in 1944. (Ullstein Picture via Getty Images)

TECHNICAL SPECIFICATIONS

THE TIGER I

The first 250 Tiger Is manufactured at Henschel's Kassel factory between August 1942 and April 1943 formed a distinct batch. The standard Tiger I was a large and angular vehicle – an unimpressed Lieutenant Otto Carius (famous for being a 'Tiger ace') described it as 'plump' – not dissimilar to the smaller Panzer IV in appearance. In total, between August 1942 and August 1944 Henschel produced 1,349 Tigers, an unimpressive total for a 24-month production run. This low figure reflected the high cost and significant time – more than twice that needed for a Panther – that had to be expended to produce a tank as large, technically complex and well engineered as the Tiger I. Yet the Tiger I's combination of lethal firepower, combat survivability and adequate mobility meant that it dominated the battlefield during 1942–44. From summer 1944 onwards, however, it began to meet its match in better-armed and better-armoured Allied rivals such as the Firefly, the Soviet IS heavy tanks and the American Pershing.

The Tiger I's undoubted battlefield prowess, however, did not mean that the design was without weaknesses. The tank's transmission was prone to breakdown if preventative maintenance was not carried out regularly; it needed a high level of general technical maintenance; ice tended to freeze on the interleaved road wheels; and it was extremely difficult to recover a disabled Tiger from the battlefield. Despite these flaws, the Tiger soon became the German tank most feared by Allied units. It continued to spearhead

OPPOSITE Taken in Tunisia in early 1943, this side view of an early Tiger shows it with its wading tube erected. Also visible is the drum-shaped original commander's turret cupola with its vision slits. (Tank Museum)

Germany's elite heavy-tank units until the latter half of 1944, after which it was increasingly replaced by the Tiger II. Nevertheless, Tiger Is continued to give sterling battlefield service, albeit in dwindling numbers, until the end of the war in early May 1945.

ARMOUR

The Tiger I possessed very thick, high-quality, homogeneous armour that ranged from the 120mm-thick mantlet armour to the 80–82mm-thick side and rear plates. The hull armour consisted of the driver's front plate (100mm at 9°), front nose plate (100mm at 25°), superstructure side plates (80mm at 0°), hull side plates (60mm at 0° vertical), tail plate (80mm at 9°), deck plates (25mm at 90°) and belly plate (25mm horizontal). This armour provided excellent battlefield survivability, even if – in comparison with later tanks – it was not well sloped.

OPPOSITE A Tiger I udergoes turret repair with the help of a crane, in the Soviet Union, summer 1943. (Bundesarchiv)

TIGER I TURRET

1. Muzzle brake
2. Armoured sleeve
3. Mantlet shield
4. Trunnion
5. Recoil cylinder
6. Articulating binocular sight
7. Breech assembly (vertical sliding breechblock)
8. Breech control
9. Commander's cupola
10. Recoil guard
11. Turret ventilator
12. Turret stowage bin
13. Used shell case holder
14. Ammunition stowage (in lockers)
15. Holder for water container
16. Turret base rotary junction
17. Turret traverse hydraulic motor
18. Gunner's seat
19. Co-axial machine-gun firing pedal
20. Gun tube

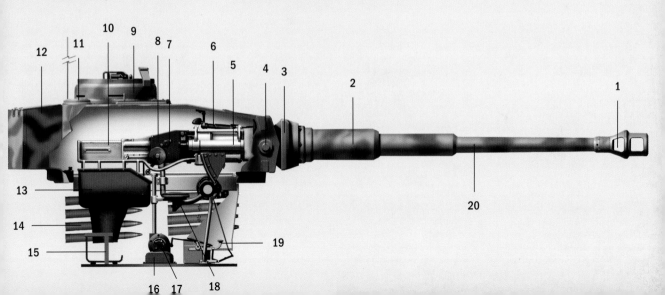

ARMAMENT

The Tiger I mounted the same potent long-barrelled 8.8cm KwK 36 L/56 main gun as the VK45.01(H) prototype. Using an armour-piercing round, this accurate gun could penetrate the side or rear armour of a Sherman V or VC Firefly at a staggering 3,500m and puncture it frontally at 1,800m. In contrast, the standard 75mm-gunned Sherman could only puncture the Tiger I's side armour at 100m, and could not even penetrate the Tiger I frontally at point-blank range. The Tiger I typically engaged an enemy tank at 800–1,200m range, although lucky kills at 2,000m were not unknown; the gun could even fire a high-explosive (HE) round 8,000m. The Tiger I carried 92 rounds for its main gun, usually a 50/50 mix of armour-piercing capped ballistic capped (APCBC) and HE rounds. Less commonly, the Tiger I carried a few armour-piercing composite rigid (APCR) and high-explosive anti-tank (HEAT) rounds.

As well as the main gun, the Tiger I had a 7.92mm MG 34 machine gun mounted coaxially to the right of the main gun, and another on the hull front for close defence. From August 1942 to June 1943 smoke grenade launchers were mounted on the turret sides to create a smokescreen to cover withdrawals. These were discontinued after an incident when small-arms fire set them off.

MOBILITY

Manned by a crew of five, the Tiger I was powered by a 642bhp Maybach HL 210 P45 petrol engine. Despite its immense size, the tank developed a satisfactory degree of mobility, obtaining top road and off-road speeds of 38kph and 20kph, respectively. Lieutenant Otto Carius of the 502nd Heavy Panzer Battalion was surprised to discover that the Tiger 'drove just like a car'. The vehicle devoured petrol, however, and thus its fuel tanks only enabled it to travel a paltry 57km off-road before it needed to refuel. It was able to move satisfactorily cross-country thanks to its unusually wide battle tracks, which ran on six interleaved layers of road wheels. When fitted with these battle tracks, however, the Tiger I was too wide to be transported on

TIGER I AMMUNITION

Sprenggranate (Sprgr) – HE round (1): A shell containing explosive filler that exploded on impact to give a large blast area; used to engage 'soft' (that is, unarmoured or lightly armoured) targets such as enemy soldiers, lorries, half-tracks etc.

Panzergranate (Pzgr) 39 – APCBC round (2): A variant of the APCR round that featured a brittle cap on top of the soft metal cap, designed to achieve good penetration against sloped armour.

Panzergranate (Pzgr) 40 – APCR round (3): A soft metal round with a small, high-density core; on impact the core was ejected at very high speed from the round, penetrating the target.

1   2   3

a standard German railway flat car. To solve this problem the Germans designed the Tiger I to use a two-track system. When in action, the Tiger used its wide battle tracks. When it needed to be transported by rail, the crew removed these tracks and the outer layer of road wheels, and fitted narrower transport tracks. The capability of the Tiger to negotiate obstacles and cross terrain was as good as or better than most German and Allied tanks, but as it was too heavy to cross many European bridges, it was fitted with wading equipment that allowed it to move submerged along the bed of a river.

PRODUCTION-RUN MODIFICATIONS

Like all tanks, the design of the Tiger I was regularly modified during the main production run. Indeed, this run can be divided into three main sub-categories: the 'early', 'mid', and 'late' production vehicles, although the transition from one category to another was by no means distinct. The 'early' Tiger Is – the first 250 tanks – have been described above. The mid-production sequence commenced in late April 1943 with vehicle 251 and continued until chassis number 824 in January 1944. All Tiger Is after vehicle 251 featured the more powerful 694bhp Maybach HL 230 P45 engine, as well as an improved transmission, which marginally boosted the vehicle's off-road performance. Next, from vehicle 391 in July 1943, Tiger Is featured a redesigned commander's cupola in the turret with armoured periscopes instead of visors. From September 1943 onwards (from vehicle 496), the design discontinued the expensive and little-used wading equipment, as an economy measure to boost delivery rates. Tiger Is completed after this date were also outfitted with *Zimmerit* paste to protect them from infantry-delivered magnetic mines.

Late-production modifications appeared during the last eight months of Tiger I production (January–

OPPOSITE A Tiger I of the 503rd Heavy Panzer Regiment, bogged down in marshy terrain. If a vehicle became stuck like this, it would have to be recovered by several halftracks, and if recovery proved impossible, the crew would have to destroy the tank. (Panzerfoto)

August 1944) and featured a varying combination of the following features. From January (vehicle 820) onwards, selected vehicles featured the multipurpose *Nahverteidigungswaffe* (Close Defence Weapon). From this time on, Tigers also began to feature resilient steel-rimmed road wheels in place of the previous rubber-tyred ones. Next, from March (chassis 920 onwards), the turret roof armour was increased from 25mm to 40mm to help protect against plunging fire. Then from around chassis 1100 in April, Tigers featured the monocular TzF 9c sight in place of the previous binocular TzF 9b one. Other minor modifications introduced during this period included the addition of stowage brackets on the turret sides to hold five spare track links and the replacement of the two hull roof-mounted headlamps by a single one fitted onto the driver's front plate.

OPPOSITE Tiger I tanks being transported on special railway lowloaders, October 1943. The wider combat tracks have been replaced with transport tracks. (Ullstein Bild via Getty Images)

THE TIGER II

ARMOUR

The Tiger II was even heavier and more thickly armoured than the Tiger I. The upper glacis plate was armoured to a thickness of 150mm, which added weight but boosted battlefield survivability. To resist penetration, tank armour needed to be hard to deflect or shatter an incoming round, but also flexible to diffuse its impact energy and retain structural integrity. Like other heavy, late-war armoured vehicles the Tiger II relied on thickness to counter most anti-tank projectiles of the period. Its hull and turret comprised RHA made from cast ingots infused with chromium and molybdenum to increase deep internal hardening and stress resistance. By compacting and consolidating the metal's microscopic grains to a consistent size and orientation, the plate was strengthened and better able to defeat an incoming round. Homogeneous armour worked best when it was the same hardness throughout, as variations promoted stress

TIGER I SPECIFICATIONS (VEHICLES 1–250)

GENERAL
Production run: August 1942–August 1944 (24 months)
Vehicles produced: 1,349
Combat weight: 56 tonnes
Crew: five (commander, gunner, loader, driver, radio/bow MG operator)

DIMENSIONS
Overall length: 8.24m
Hull length: 6.20m
Width (with battle tracks): 3.73m
Height: 2.86m

ARMOUR (THICKNESS AT DEGREES FROM VERTICAL)
Hull front: 100mm at 66–80°
Hull sides: 60–80mm at 90°
Hull rear: 82mm at 82°
Hull roof: 25mm at 90°
Turret front: 100–120mm at 80–90°
Turret sides: 80mm at 90°
Turret rear: 80mm at 90°
Turret roof: 26mm at 0–9°

ARMAMENT
Main gun: 1 x 8.8cm KwK 36 L/56
Secondary: 2 x 7.92mm MG 34 (1 co-axial in turret, 1 hull front); 2 x treble smoke dischargers (turret sides)
Main gun rate of fire: 15rpm

AMMUNITION STOWAGE
Main: 92 rounds (typically 50 per cent Pzgr 39 APCBC, 50 per cent Sprgr L/4.5 HE; also a few Pzgr 40 APCR, Gr 39 HL HEAT)
Secondary: 3,900–5,100 rounds

COMMUNICATIONS
FuG 5 ultra-short-wave transmitter/receiver; intercom

MOTIVE POWER
Engine: Maybach HL 210 P45 12-cylinder 21-litre petrol engine
Power: 642 metric bhp at 3,000rpm
Fuel capacity: 534 litres
Power-to-weight ratio: 11.5hp/tonne

PERFORMANCE
Ground pressure: $1.05 kg/cm^2$
Maximum road speed: 38kph
Maximum cross-country speed: 20kph
Operational range (road): 100km
Operational range (cross-country): 57km
Fuel consumption (road): 5–5.3 litres/km
Fuel consumption (cross-country): 9–9.3 litres/km

concentration boundaries and weakened its ballistic resistance. With the vehicle's glacis and mantlet 150mm and 180mm thick respectively, achieving such consistency was not easy.

As the production process was consistently hampered by Allied bombing, the tempering process of heating the raw metal to 800°C, cooling it in water, reheating it at a lower temperature and cooling it again could not always be done to the accuracy required to produce the desired ductility of alloyed steel armour plate. As a result of this 'scale effect', a crystalline microstructure (collectively called bainite) could form internally, which increased hardness and the potential for cracking on impact.

Subsequently, an impacting projectile's shock wave would be likely to produce an internal showering of sharp metal flakes known as spall. The Germans tended to use the Brinell scale to determine armour's

OPPOSITE Tiger I tanks in production at Henschel's factory, 1944. (Bundesarchiv)

hardness, which on the Tiger II's glacis and hull sides were BHN 220–265 (150mm) and 275–340 (80mm), respectively. With late-war stockpiles of molybdenum, nickel and manganese dwindling as the war progressed, vanadium was used as a grain-growth inhibitor to improve the RHA's toughness.

As a smaller profile reduced the chance of being hit, the Tiger II's turret, probably the most exposed part during combat, was tapered at the front, and backed up with 180mm of face armour and a dense, curved *Saukopf* mantlet. This meant that when the barrel was pointed at an adversary, the turret would either defeat front-on shots owing to that area's great thickness or deflect those striking the sides owing to the great angle. A second benefit of using hard, thick armour was that high-speed incoming rounds often fell into a 'shatter gap' where they simply disintegrated on striking the vehicle. Sloped armour also increased the plate's effective thickness. Combined, these aspects translated into the vehicle's frontal armour being essentially impenetrable to existing Allied guns,

while side plate armour proved adequate when fighting at the commonly long ranges afforded by the main gun.

ARMAMENT

For the Tiger II, Krupp and Rheinmetall-Borsig produced two prototypes of the new 8.8cm KwK 43 L/71 gun, with the first being an entirely new design and the second simply a reworked FlaK 41 L/74. As Krupp's version was shorter, possessed a muzzle brake and used shorter, more easily stored projectiles, it was deemed superior and accepted for production. As an internally mounted variation of the PaK 43 anti-tank gun, it was initially developed with a mono-block barrel and mated to Porsche's 'pre-production' turret. Considerable stress from firing high-velocity rounds, however, necessitated a change to a two-piece weapon, which eased construction and the ability to change barrels. A falling-wedge breechblock ejected spent shell casings and remained open for another round, and because of the main gun's large size, a muzzle brake

TIGER II SPECIFICATIONS (VEHICLES 269–72; 350–55; 362–90)

GENERAL

Production run: November 1943–March 1945 (17 months)
Vehicles produced: 492 (including three prototypes)
Combat weight: 69.8 tonnes (with 'series' turret)
Crew: five (commander, gunner, loader, driver, radio/bow MG operator)

DIMENSIONS

Length (hull/overall): 7.620m/10.286m
Width (without aprons/with aprons): 3.650m/3.755m
Height: 3.09m

ARMOUR (THICKNESS AT DEGREES FROM VERTICAL)

Glacis (upper/lower): 150mm/100mm at 50°
Hull side (upper/lower): 80mm at 25°/80mm at 0°
Hull rear: 80mm at 30°
Hull roof: 40mm at 90°
Hull bottom (front/rear): 40mm/25mm at 90°
Turret face: 180mm at 10°
Turret mantlet: 150mm (*Saukopf*)
Turret side: 80mm at 21°
Turret rear: 80mm at 20°
Turret roof: 40mm at 78°
Cupola side: 150mm at 0°

ARMAMENT

Main gun: 8.8cm KwK 43 L/71 (22 turret/64 hull) (typically 50% Pzgr 39/43 APCBC and 50% Sprgr 43 HE)
Secondary: 2 × 7.92mm MG 34 (coaxial; bow); additional 7.92mm MG (anti-aircraft) (5,850 rounds)
Main gun rate of fire: 5–8rpm

COMMUNICATIONS

Internal: *Bordsprechanlage* B intercom
External: FuG 5 10-watt transmitter/ultra-short-wave receiver.

MOTIVE POWER

Engine: Maybach HL 230 P30 12-cylinder (water-cooled) 23 litres (petrol)
Power-to-weight: 600hp (sustained) at 2,500rpm; 700hp (max.) at 3,000rpm (10hp/tonne)
Transmission: Maybach Olvar 40 12 16 B; eight forward, four reverse gears
Fuel capacity: 860 litres in seven tanks

PERFORMANCE

Ground pressure (hard/soft): 1.03 kg/cm^2 / 0.76kg/cm^2
Maximum speed (road/cross-country): 41.5kph/20kph
Operational range (road/cross-country): 170km/120km
Fuel consumption (road/cross-country): 5.1 litres/km / 7.2 litres/km

TIGER II 'SERIES PRODUCTION' TURRET

1. Recoil cylinder
2. Coaxial MG 34 (7.92mm)
3. Loader's forward vision port
4. Turret ventilator
5. Localised support port (smoke, flares etc.)
6. Manual turret rotation wheel
7. Loader's seat
8. Ammunition transfer support roller
9. Falling wedge breech
10. Commander's seat
11. Gunner's seat
12. Traverse and elevation wheels
13. Commander's cupola
14. Ring mount for 7.92mm MG (anti-aircraft)
15. TzF 9d monocular sight
16. Ammunition storage

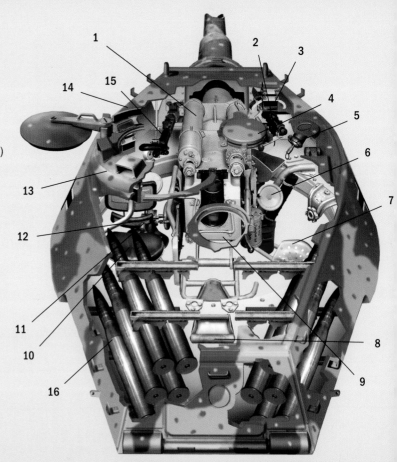

was installed both to vent unwanted propellant gases and to reduce recoil.

As a secondary armament, like its predecessor, the Tiger II was fitted with two 7.92mm machine guns.

MOBILITY

To avoid production delays and to maximise vehicle hardware interchangeability, it was decided to use the HL 230 P30 engine. Built by Maybach, Auto Union (four automobile manufacturers, including Audi) and Daimler-Benz, it was used in other heavy German armoured vehicles including the medium 45-tonne Panther and late-model Tiger I. Because of the Tiger II's additional weight, a transverse torsion-bar suspension system comprising nine load-carrying axles per side was incorporated. This provided independent wheel movement in the vertical, increased stiffness in turns, helped retain stability over rough terrain and allowed a theoretical maximum speed of 41.5kph over hard, level surfaces, although a much lower pace was recommended during general operation. In contrast with the Tiger I's interleaved road wheels, its successor incorporated a twin steel-rimmed, rubber-cushioned type, which improved maintenance and cold-weather operation as ice and snow were less likely to impede rotation. Changing gears was surprisingly easy for a front driving sprocket that provided power to the 'shoe' and 'connector link' style continuous tracks, which were tensioned by a rear idler, and controlled via power steering.

COMMUNICATIONS

Internally, crew communication was conducted through the *Bordsprechanlage* B intercom. As the standard intercom box installed on vehicles from the Panzer III to the Tiger II, it comprised an audio frequency amplifier for volume control. An FuG 5 10-watt transmitter/ultra-short-wave receiver was used for external communication to ranges of 6km and 4km, respectively. The associated 2m hollow sheet-steel rod antenna was mounted on a rubber base for additional flexibility when travelling through heavy foliage or under low obstructions such as bridges.

TIGER II AMMUNITION

The **Pzgr 39/43 (1)**, an APCBC/HE-T (high explosive-tracer) round, was the Tiger II's primary anti-tank round. Designed to handle the high internal barrel pressures within the KwK 43 L/71 gun, it possessed a tracer and a second driving band for added stability and accuracy over its Pzgr 39 predecessor. Its hard shell was capped by softer metal to minimise disintegration from high-velocity strikes.

The **Sprgr 43 (2)**, an HE round, was used against unarmoured vehicles, infantry and static defensive positions. The projectile had no tracer, and except for a second driving band it was the same as the older Sprgr version.

The **Gr 39/43 HL (3)** was a HEAT round that relied on a shaped-charge to penetrate armoured vehicles, and which could also be used by other 8.8cm guns as indicated by the text on the cartridge. As only about 7,000 shaped-charge Gr 39/43 HL rounds were produced, their use was uncommon.

The **Pzgr 40/43 (4)**, an HVAP-T (hyper-velocity armour piercing-tracer) round, was to be used against the thickest enemy armour. The limited availability of tungsten after 1943 meant that this armour-piercing projectile was also produced with steel- or iron-core expedients. The round was a kinetic penetrator and had a smaller explosive charge than the Pzgr 39/43. Because of its lighter weight, the shell was affected by wind resistance and had decreased accuracy. Only about 5,800 Pzgr 40/43s were made.

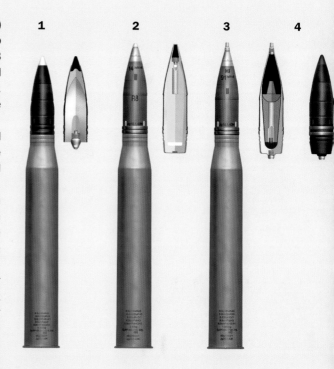

1   2   3   4

ABOVE The thick armour plate of a Tiger II, used after the war for test firings by the US military. The 150mm armour has not been penetrated by HVAP shots from 90 and 105mm guns, which at that time were the most powerful US-manufactured anti-tank weapons in service. (NARA)

CREW TRAINING AND ORGANISATION

TRAINING

All recruits for the German Army and Waffen-SS began their military service with a programme of basic training as infantrymen. Only when this had been completed would the recruits apply to undertake further training – either as other ranks or as aspirant officers – with the specialised branch of their choice: Panzer troops, mountain troops, engineers, signallers etc. Only a proportion of those recruits and aspirant officers who put themselves forward for the popular branch of the Panzer forces were accepted for specialised training. The Panzer arm invariably selected only those who had excelled in basic training. To fill the role of tank commander, the branch sought to recruit exceptional leaders who could swiftly size up a complex tactical situation and execute a timely decision based upon this appraisal. The Panzer branch also required experienced mechanics and highly skilled technicians to serve as tank drivers, gunners and wireless operators. The tank loader was probably the least technically qualified member of the crew, but this physically demanding role was just as vital to the performance of the tank as the others. In practice, some Tiger commander roles were filled by accomplished former gunners or drivers who had

OPPOSITE A Tiger I crew from the 502nd Heavy Tank Regiment dressed in the standard black armoured corps uniform. The factory-new tank is painted in standard dark yellow and is fitted with six smoke grenade dischargers on the turret. A further six are mounted on the top of the hull. (Panzerfoto)

completed specialist refresher courses. It was less common for a loader to be promoted to command a tank, although this is precisely what occurred with Otto Carius, who went on to become one of Germany's leading Tiger aces.

Specialised Panzer-arm training was undertaken at one of the many dedicated armoured training schools located within the Reich. Personnel received extensive training in the tank-crew role for which they had been selected, as well as more superficial training in the other crew roles. Such training also sought to instil within a particular tank crew a smooth and efficient interaction between the various team members. Nowhere was this interaction more crucial than in the interplay between commander, gunner and loader in the drills performed for engaging enemy tanks. The crews honed these skills by regularly conducting live

OPPOSITE The radio operator of a Tiger crew prepares sandwiches for himself and his comrades, Eastern Front, August 1943. (Photo by Heinrich Hoffmann/Ullstein Bild via Getty Images)

firing on the ranges and proving grounds located within the Reich; around 20 of the earliest Tiger Is had been allocated to the training schools for this purpose.

The basic German tank engagement drill ran as follows. Once a tank commander had spotted an enemy tank, he would indicate the bearing and order the crew to engage it. If necessary, the driver would move the Tiger to a better firing position. Then, looking through his gunsight, the gunner would calculate the range and lay the main gun onto the target, compensating for expected trajectory disturbances caused by strong cross-winds or the spinning of the round in flight. In the meantime, the loader had manhandled a long Tiger APCBC round into the gun's breech mechanism. Unless the target was at close range, most first anti-tank shots from a Tiger were regarded merely as an acquisition shot, expected to be more likely to land near the target rather than hit it. The gunner then corrected the range (by bracketing the range up or down in increments of 50m–200m), and/or the direction (by aiming off the target to allow for wind and round deflection in flight). With these compensations made, the crew then fired a second round. This shot – termed firing for effect – was expected to hit (and hopefully destroy) the target. German training taught crews to expect to be the first tank to hit an enemy target, rather than to be the first tank to fire in any given engagement.

In each military district the Commander of Panzer Troops controlled at least one school along with a host of Panzer training units, where basic gunnery training took place. In addition to these basic gunnery drills, the training done at these schools covered the whole gamut of professional knowledge. The recruits received instruction and practical exercises in the science of ballistics, in the various drills associated with vehicle maintenance and effective use of the tank's communications devices. In addition, personnel received instruction in combat tactics and the tactics of Tiger tanks cooperating with other combat arms, notably the Panzergrenadiers, the anti-tank troops and the artillery. This all-arms capability was tested

OTTO CARIUS

One of the leading exponents of the art of effective combat in the Tiger was Lieutenant Otto Carius, who as reward for his total of over 150 enemy 'kills' received the Oakleaves to the coveted Knight's Cross. Born on 27 May 1922 at Zweibrücken, Carius finally managed to voluntarily join the German Army in May 1940, having been previously turned down twice on the grounds that he was underweight. After basic training as an infantryman, Carius put his name forward for the much sought-after armoured forces. Subsequently, he served as a loader in a Panzer 38(t) of the 1st Company, 21st Panzer Regiment.

Immediately on completing its training, the German high command committed this regiment, as part of its parent formation 20th Panzer Division, to Operation *Barbarossa* – the 1941 invasion of the Soviet Union. During the *Barbarossa* campaign, the now Sergeant Carius was wounded in action, for which he was awarded the Wound Badge in Black. In late 1942 Carius underwent officer training before being posted to the 502nd Heavy Panzer Battalion in April 1943. As a commander of a Tiger I in the battalion's second company, Carius served during 1943–44 on the northern sector of the Eastern Front. It was during these battles that Carius' mastery of the Tiger

(Bundesarchiv)

became evident. On 22 July 1944, Carius' tank, plus another Tiger, launched a bold counter-attack on a Soviet armoured spearhead that had advanced to the village of Malinava, north of Daugavpils in Latvia. Catching the enemy by surprise, the accurate fire of Carius' Tiger dispatched some 16 T-34s and one new IS heavy tank in a matter of 20 minutes. This stunning success ranks alongside Michael Wittmann's June 1944 victory at Villers-Bocage as probably the most impressive Tiger action of the entire war.

In August 1944, Carius took command of the 2nd Company of the newly forming 512th Heavy Anti-Tank Battalion, which was to be equipped with the monstrous 70-tonne Jagdtiger tank destroyer. By early 1945, this unit was still in training with its new vehicles at Döllersheim near Vienna, as the Western Allies successfully advanced towards the Rhine. On 8 March 1945, the desperate German high command felt compelled to commit the part-trained battalion to action on the Western Front near Siegburg. Despite Carius' tactical abilities, his 2nd Company could not prevent American forces from overwhelming the flimsy German defensive screen thrown up along the eastern bank of the Rhine.

Indeed, by mid-April the battalion had been surrounded – along with most of Army Group B – in the Ruhr. Carius' unit surrendered to American forces alongside some 300,000 other German troops. Whether the mighty Jagdtiger would have withstood the Firefly's potent gun remains uncertain, as Carius' company only saw service against the Americans, who did not generally use 17-pounder-equipped Shermans. After his release from American captivity, Carius went on to run a pharmacy named, rather appositely, *Der Tiger Apotheke*, and died in January 2015 at the ripe old age of 92.

ABOVE A Jagdtiger bogged down in soft ground, being inspected by US troops. (US Signal Corps)

exhaustively during the final phases of specialised training, when various tank crews practised operating as coherent tactical units – troops or companies – in a series of exacting field exercises. Personnel also participated in such exercises if they joined a Panzer regiment that was working up, rebuilding or re-equipping. Such exercises took place at armoured manoeuvre areas, like that at Putlos, in northern Germany, and Senne, near Paderborn. Armoured demonstration units were often attached to these grounds, where the experienced, high-calibre combat veterans employed in such units demonstrated the correct tactics to be employed by a Tiger unit.

DAILY LIFE

Daily life for the Tiger crews was essentially similar to that of any tank crew in World War II. The tanker's world was a small one. The nucleus of his existence was the vehicle's crew – a small group of four or five individuals. With so much of the tanker's life spent in a cramped metal container, and with the shared experiences of mortal danger and adrenalin rush of combat, tank crews soon became tightly knit cohesive communities.

That is not to say that the commander, whatever his rank, was not the centre of the crew's world, for the crew's fate rested upon the speed of the commander's reactions and the correctness of his decisions. Yet, beyond this, the survival of the crew also depended on how well each crewman performed his own specialised role. No matter how quickly a keen-eyed commander spotted an enemy tank and ordered it to be engaged, the ability of the tank to hit its opponent first rested on the driving skills of the driver, the strength and dexterity of the loader and the marksmanship of the gunner; even the wireless operator played a part if supporting vehicles were required on the scene. Therefore, a particularly strong sense of functional interdependence developed within a tank crew who were only as strong as the weakest man amongst them.

Tank crews also developed a very personal relationship with their vehicle, in which they spent so

TIGERFIBEL: THE TANK GUIDE FOR TIGER CREWS

The principal doctrinal guide for Tiger crews was the *Tigerfibel*, or 'Tiger manual', which was accepted as Field Service Regulation D656/27 by Inspector-General of Panzer Troops Colonel-General Heinz Guderian on 1 August 1943. The manual was unlike other German service regulations, which were famed for their dry technical prose, and instead was an informal and humorous practical guide to the use of the Tiger in combat. The book used amusing rhymes, mottoes, jokes and cartoons to convey the wealth of common-sense ideas it contained within its covers. The guide went through the duties of each crewman in turn, presenting regulations, useful practices and the 'tricks of the trade'. It implored gunners to 'think before you shoot' because 'for each round that you fire, your father has paid 100RM in taxes [and] your mother has worked a week in the factory'. The *Tigerfibel* likened effective driving of the Tiger to a Viennese waltz, and admonished drivers from crashing through buildings, as the dust clogged the air filters and radiator, causing the engine to overheat. The tactical discussion of how to engage enemy tanks presented a list of anti-heroes that featured characters such as generals Lee and Sherman, as well as King Voroshilov I.

ABOVE Cleaning the gun barrel required four men to push and pull the stiff brush through the bore in order to remove metal or powder residues from the rifling. (Panzerfoto)

much time. Crews soon grew accustomed to the nuances of the rumblings of the engine and the chink of changing gears. After all, their lives depended on such things working effectively. To many Tiger crews, therefore, the tank itself became a living thing that represented the sixth member of the crew. Often the tank was the only female in the group, identified with popular names such as 'Irma' and 'Brunhilde'.

Like most tankers, Tiger crews spent much of their time out of combat, engaged in routine maintenance work. This was particularly true of the Tiger, which was a complex-engineered vehicle that needed constant attention to ensure that it functioned properly. Crews would spend considerable time cleaning and inspecting the vehicle, checking for signs of leaks, scrutinising the tracks for signs of damage, assessing whether the road wheels were adequately lubricated, checking that items had been safely stowed, adjusting camouflage and cleaning the gun barrel. Though these tasks were tedious in the extreme, most crews fully understood their importance.

Tiger commanders were trained to engage enemy anti-tank guns or tanks at 800–1,200m range. The powerful cannon could, however, obtain a hit at distances up to and beyond 2,000m and – with some good fortune – disable or even knock out the enemy tank at such long ranges. Their training and combat experiences led Tiger commanders to strive to attack the enemy's more vulnerable side or rear armour, whilst protecting their own less well-protected side and rear plates from enemy fire. This could be most readily achieved, if the battlefield situation was conducive, by assuming a concealed ambush position and knocking out the unsuspecting enemy from the flank or rear.

A tank commander's ability to freely observe the battlefield was a crucial factor in determining a successful outcome to an encounter with an enemy tank. However, with the commander safely battened down in the turret with his cupola hatch closed, he could only see various parts of the battlefield – a series of compass points – through his armoured visors or periscopes. Experienced commanders like Wittmann

soon learned the value of going into action with their heads sticking up through the open turret cupola providing a good field of vision for observation of the battlefield. Panzer training stressed the importance of being the first to get a round off in an engagement, or – more critically – to be the first to hit the enemy.

Although service in an armoured vehicle would seemingly present a reduced physical risk compared with infantry and artillery units that were not protected against bullets, shrapnel and weather, Tiger crewmen faced their own hazards. While the infantry suffered higher physical casualties, tankers were afflicted with a greater incidence of mental disorders related to working and fighting in a hot, claustrophobic environment where direct contact with the outside world was limited. Constant vibration caused knee and back problems,

oedema (fluid pooling), muscle atrophy and radiculitis (spinal-nerve inflammation). Explosions from anti-tank mines or forceful, non-penetrating projectile impacts could cause blunt trauma and shower the interior with spall. Carbon monoxide build-up was especially problematic, especially when the vehicle was 'buttoned up' with its hatches closed. Noise was an ever-present problem during operation, where those exposed to constant noise exceeding 85dB suffered hearing damage. With the firing of main guns and movement often producing 140dB and between 120dB and 200dB respectively, effective communication and target detection was often hampered due to crew disorientation.

ORGANISATION

When the Tiger I tank first entered German service on the Eastern Front in August 1942, it was made even more formidable as an individual battlefield weapon by being grouped into independent heavy armour battalions of up to 45 vehicles. Originally intended to add an offensive, heavy-combat element during

OPPOSITE July 1943, battle of Kursk. A Tiger I tank advancing toward a burning village. The commander here seems to be using the view from the cupola to gain better visibility over the battlefield. (Photo by Weltbild/Ullstein Bild via Getty Images)

TIGER I, 506TH HEAVY PANZER BATTALION, RUSSIA, 1943

The 506th Heavy Panzer Battalion, formed in July 1943, was the first to adopt the new organisation, with a complement of 45 Tigers from the very beginning. Most of these Tigers had the redesigned turret with new features including the *Prismenspiegelkuppel* (commander's cupola with periscopes and swivel hatch). Originally delivered from the factory with a base coat of dark yellow (RAL 7028), the units were issued with water-based whitewash for camouflage during the winter, which was to be cleaned off during the spring thaw.

During this period, the Tigers of the 506th used tactical numbers 1 through to 14 stencilled on the turret sides to designate the position of a Tiger within the company. The colour of the number and the 'W' in the unit badge was used to identify the company. The emblem for the 506th was a Tiger holding a shield emblazoned with a white cross astride a 'W', stencilled on the rear of the turret stowage bin.

German breakthroughs, after mid-1943 these formations were increasingly used in defensive 'fire brigade' or mobile reserve roles to help shore up threatened sectors. With exceptions made for research, training and allocation as company-sized complements to select Army and Waffen-SS divisions, Tiger I and II units were subordinated to corps and army-level commands. Throughout the war, ten heavy Panzer battalions were created for the Army, which operated in North Africa, Italy, north-west Europe and on the Eastern Front. After April 1943 four more were to be organised for the Waffen-SS, with three coming to fruition as the 101st, 102nd and 103rd Heavy SS-Panzer battalions serving under I, II and III (Germanic) SS-Panzer Corps, respectively. Finally, there were also four army heavy radio-controlled tank companies, in which Tigers operated alongside remote-controlled demolition vehicles. The Western Army (*Westheer*) only ever deployed one army and two SS Tiger battalions, plus one Tiger company, to the Normandy front line.

Initially, heavy Panzer battalions officially comprised 20 Tiger Is and 16 Panzer IIIs that were organised into two armoured companies, each of four platoons; two of each vehicle were allocated per platoon, plus a Tiger I for each company commander and a pair for battalion command. As Tiger I and II production increased, the 'medium' vehicles were eliminated from the roster.

TACTICS

Panzer doctrine stressed offensive (almost to the exclusion of defensive) combat, as evidenced by the 1941 Army Service Regulation (*Heeres-Dienstvorschrift*) 470/7. Armour regiments and battalions were doctrinally to employ one of three offensive actions. Firstly, in a *Vorbut* (meeting

OPPOSITE Two whitewashed mid-production Tiger I tanks carefully camouflaged at the edge of a wood, ready for immediate action. The near tank carries the tactical number 1 and is coated with *Zimmerit*, anti-magnetic paste. (Panzerfoto)

engagement), an advance force, generally at least company-sized, was employed for taking an enemy by surprise so as to gain key terrain or a similar objective. Alternatively, *Sofortangriffe* (quick attacks), often conducted using the *Breitkeil* (broad wedge) formation (a reverse wedge with two platoons forward and the remaining platoon providing flank support as required), were used when supporting forces were not readily available and immediate action was needed. Finally, an *Angriff nach Vorbereitung* (deliberate attack) could be conducted as a complete unit against prepared defences.

As a result of the German Army's effective training and discipline, commanders were generally given a considerable degree of flexibility in carrying out tactical and small unit operations. Given little more than the mission and the leader's intention, commanders could conduct an operation and adjust to battlefield situations as they occurred. This was perhaps best exemplified in the *Panzerwaffe* where timely intelligence, security, march discipline and communications were especially important elements in achieving victory on the battlefield.

When the Tiger I made its combat debut in November 1942, its battlefield tactics and the operations of heavy armour battalions were largely improvised by the crewmen based on their personal experiences. By the time the Tiger II was first fielded in early 1944, a more established doctrine had been developed to address specific battlefield eventualities in a unified manner. Officially, four-vehicle armour platoons were to deploy in either a *Linie* (line – section leader/vehicle/platoon leader), a *Reihe* (row – platoon leader/vehicle followed by section leader/vehicle), a *Doppelreihe* (double row) or a *Keil* (wedge), but in practice terrain, the situation and the commander's experience meant these 'parade' formations were seldom used.

Tiger crews focused on achieving surprise and executing decisions quickly. Once fields of fire were established, terrain was to be exploited for protection and concealment. When facing overwhelming numbers, tankers were to scatter and regroup to a more advantageous position. On spotting hostile tanks, the Tigers were to halt and get ready to engage them by surprise, estimating the enemy's reaction before launching an attack. The crew should hold fire as long as possible, and, if possible, the enemy tanks should not be engaged by a single tank so as to apply maximum firepower. In practice, however, deviations were unavoidable, and at the commander's discretion. To facilitate these, emphasis was laid on the importance of rapid communications between the battalion and its parent command, as such formations required as much time as possible in which to get organised and allocated to a threatened sector. There was also a heavy reliance on armoured engineers to strengthen bridges and clear minefields ahead of the heavy tanks.

BELOW A Tiger I from the 503rd Heavy Panzer Battalion is towed to the workshop. (Panzerfoto)

ABOVE A Tiger from the 508th Heavy Panzer Battalion destroyed by US forces in 1944. The impact of the armour-piercing (AP) round has chipped off large areas of the *Zimmerit* anti-magnetic coating. (US Signal Corps)

TIGERS VS ALLIED ARMOUR

The Tiger's heavy armour and powerful gun made it a fearsome opponent, spurring the Allied powers to produce tanks which could contend with it in the field. For the British, this was the Sherman Firefly, an up-gunned variant of the standard American-designed M4 Sherman medium tank, whose 17-pounder gun could penetrate the Tiger's armour. Later in the war, Soviet engineers produced the IS-2 to similar effect. Two encounters between a Tiger I and II and Allied armour – one in Normandy in 1944 and one in East Prussia in 1945 – are described in this chapter.

TIGER I AND SHERMAN FIREFLY – NORMANDY, 1944

THE STRATEGIC SITUATION

During spring 1944, the forces of General Bernard Montgomery's British 21st Army Group trained relentlessly for the Allied D-Day landings on the coast of German-occupied Normandy, finally initiated on 6 June. Both sides knew that the outcome of these landings – and the ensuing battle for Normandy – held monumental significance for the course of the war. It was not surprising, therefore, that during this period the first Sherman Fireflies to be completed were rushed to the British and Canadian armoured units then finalising their preparations for D-Day. These units, which included the 1st Northamptonshire Yeomanry, held high hopes that the Firefly would enable them to defeat the enemy's most potent tanks, including the Tiger I. Senior German commanders in turn held equally high expectations for what the Tiger would contribute to the imminent struggle, hoping that it would spearhead the defensive and the counter-offensive actions required to halt and even reverse the Allied invasion. This was a particularly tall order, as the German Western Army typically deployed fewer than 80 operational Tiger Is. Nevertheless, battlefield

(Bundesarchiv)

MICHAEL WITTMANN

Wittmann's gunner: *They are behaving as if they've won the war already.*
Wittmann: *We're going to prove them wrong.*
(Villers-Bocage, Normandy, 13 June 1944)

Born a farmer's son on 22 April 1914, Michael Wittmann rose to become one of Germany's leading Tiger aces. After service in the German Army as a private during 1934–36, Wittmann enlisted in the elite 1st SS-Panzer Division Leibstandarte SS Adolf Hitler, served in the 1939 Polish campaign, and led a StuG III assault gun platoon during the spring 1940 Balkan war. Next, he participated in the invasion of the Soviet Union, receiving an Iron Cross First Class and promotion to *SS-Oberscharführer* (SS-Senior Company Leader) for his outstanding performance as a destroyer of Soviet tanks. After officer training, now *SS-Untersturmführer* (SS-Under Storm Leader) Wittmann rejoined the Leibstandarte in December 1942. While serving with the division's Tiger I-equipped 13th Heavy Company, he again performed well during the July 1943 battle of Kursk, thanks to his careful planning of actions and the 'unshakable calm' he maintained during combat. This company then formed the nucleus of the newly raised

101st Heavy SS-Panzer Battalion, in which Wittmann continued to serve until his death on 8 August 1944.

During January 1944, the newly promoted SS-*Obersturmführer* (SS-Senior Storm Leader) Wittmann received the Knight's Cross, and then the Oakleaves to this coveted award, for his tally of over 90 enemy kills. By the time he assumed command of the 101st Battalion's 2nd Company in March, Wittmann had also married Hildegard Burmester, with his gunner, Bobby Woll, as his best man. Joseph Goebbels' propaganda machine now seized upon Wittmann's exploits and transformed this modest yet determined officer into a national hero.

On 13 June 1944, Wittmann joined combat in Normandy with the 101st Battalion. That day his Tigers inflicted a bloody repulse on the British 7th Armoured Division at Villers-Bocage. Wittmann's Tiger I was on reconnaissance when he observed the spearhead tanks of the British 22nd Armoured Brigade advancing through the hazy daylight near Villers-Bocage. Wittmann's Tiger moved west behind the column while four other Tigers moved east to attack its spearhead. Catching the column by surprise, the Tigers poured fire into it, leaving around 20 enemy vehicles burning furiously. So far the Tigers had triumphed, but the second half of the action did not go as well for them. By the time that Wittmann, possibly supported by two other Tiger Is, headed west into Villers-Bocage, the 22nd Armoured Brigade had established an effective defensive position. One Sherman Firefly, three Churchill tanks and a 6-pounder anti-tank gun had deployed in the town's side streets, ready to ambush the Tigers with close-range fire against their more vulnerable side armour. As the Panzers moved through the main street, the anti-tank gun engaged and disabled Wittmann's Tiger, forcing the crew to flee on foot. Ultimately that day the Germans lost at least four Tiger Is in the actions that raged all day around Villers-Bocage, while the Allies lost at least 10 tanks and around 20 other vehicles.

With this feat behind him, Wittmann served in the bitter defensive stands the Germans enacted in and around Caen during July. Yet on 8 August – by which time the now SS-*Hauptsturmführer* (SS-Head Storm Leader) Wittmann had claimed 139 combat kills – the Panzer ace met a warrior's end during a desperate counter-attack launched against numerically superior Allied forces.

events within a week of the landings showed that these grandiose German expectations were not unrealistic. On 13 June, in a famous incident, a few Tiger Is led by Michael Wittmann mauled the British armoured brigade that had thrust audaciously south to Villers-Bocage.

Although such isolated examples of successful German counter-attacks were insufficient to stop the gradual expansion of the Allied beachhead in Normandy, the Tiger I's defensive prowess did help the Germans slow the Allied advance to a crawl. By mid-July the Tiger I's capabilities had helped turn the Normandy campaign into a bloody attritional war of materiel. These hard-fought battles again demonstrated the vulnerability of the Sherman (and thus the Firefly) to German tank fire, as well as the Tiger I's virtual invulnerability to 75mm-gunned Sherman fire. On 18 July, during Operation *Goodwood*, for example, 11th Armoured Division lost 21 of its 34 Fireflies to enemy tank and anti-tank fire in just one day. Despite such setbacks, on numerous occasions Sherman and Firefly crews bravely engaged the Tiger Is, with many of them paying the ultimate price for their gallantry. An unofficial dictum soon sprung up in British armoured units: if a Tiger appears, send out a troop of four Shermans (with its single Firefly) to destroy the Panzer, and expect only one to come home. Understandably, after such painful combat experiences some Allied tank crews became so concerned about the Tiger I's capabilities that 'Tiger-phobia' became evident. One brigadier recorded an extreme manifestation of this phobia, when a solitary Tiger I 'fired for one hour … [and] then drove off unmolested … [because] not one tank went out to engage it'.

Despite the disproportionate damage done by the few Tiger Is deployed in Normandy, the combination of Allied numerical superiority, offensive determination and the Firefly's lethal gun gradually wore down the German Western Army's powers of resistance, while Allied tank crews' confidence increased. For the bitter armoured battles waged in Normandy showed that the Firefly's potent 17-pounder gun could indeed take out

ABOVE A destroyed British Cromwell tank and M4 Sherman on the road at Villers-Bocage, June 1944. (Bundesarchiv)

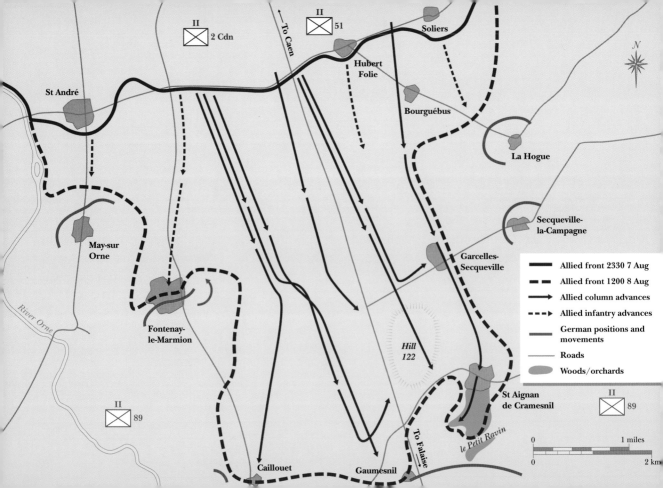

II 2 Cdn

To Caen

II 51

Soliers

Hubert Folie

St André

Bourguébus

La Hogue

May-sur Orne

Secqueville-la-Campagne

River Orne

Garcelles-Secqueville

Fontenay-le-Marmion

Hill 122

St Aignan de Cramesnil

II 89

To Falaise

le Petit Ravin

Caillouet

Gaumesnil

	Allied front 2330 7 Aug
	Allied front 1200 8 Aug
	Allied column advances
	Allied infantry advances
	German positions and movements
	Roads
	Woods/orchards

II 89

0 | | 1 miles
0 | | 2 km

even the much-feared Tiger I. During a 96-hour period of the late-June *Epsom* offensive, British 11th Armoured Division Fireflies knocked out or disabled five Tiger Is in a series of bitter encounters. Aided by the Firefly's firepower, the repeated Allied attacks finally bore fruit in late July when Lieutenant-General Omar Bradley's First US Army *Cobra* offensive broke through the German front around Saint-Lô. Just prior to this, Montgomery's *Goodwood* offensive had advanced the Anglo-Canadian front to the Bourguébus Ridge, from where the *Totalize* offensive commenced on 7–8 August. Between 31 July and 6 August, American forces – having passed the bottom corner of the Cotentin Peninsula and with the German line shattered – raced west into Brittany, south towards the Loire and east towards the Seine.

Hitler now blundered strategically by ordering the German Western Army to counter-attack the narrow corridor located behind the American breakout. Consequently, during 6–7 August, German armour attacked west towards Mortain. This ill-advised attack predictably failed and in so doing sucked German forces further west, pulling them deeper into a large encirclement that was beginning to form in the Mortain–Argentan area. For by 8 August – by which time *Totalize* had commenced – American forces had raced south-east to capture Le Mans, deep into the German rear. Meanwhile, the western part of the Allied front, manned by Montgomery's two Anglo-Canadian armies, had remained relatively static, from the coast to the Bourguébus Ridge and thence south-west to Vire, where the American sector began. By 8 August, therefore, it now seemed possible that if the infant *Totalize* offensive could secure Falaise, its forces might then be able to link up around Argentan with the north-westerly American advance beyond Le Mans. In so doing, the Allies would have encircled substantial German forces in a giant pocket.

OPERATION *TOTALIZE*

Operation *Totalize*, spearheaded by Lieutenant-General Guy Simonds' II Canadian Corps, aimed to advance

24km south from the Bourguébus Ridge to secure the high ground that dominated the town of Falaise.

Simonds' five divisions and two independent brigades would attack the defences manned by the 89th Infantry Division of I (Germanic) SS-Panzer Corps, while the 12th SS-Panzer Division Hitlerjugend (Hitler Youth) held reserve positions further south. During the offensive's first phase, two infantry divisions would attack south astride the main Caen–Falaise road in an initial surprise nighttime break-in operation, aided by heavy bombers. Seven mobile columns would spearhead the attack by audaciously infiltrating between the forward defensive localities (FDLs) to seize objectives deep in the German rear. Simultaneously, infantry would attack the FDLs the columns had bypassed.

Aided by surprise, during the night of 7/8 August this first phase of *Totalize* went extremely well. By mid-morning on the 8th, Simonds' forces had secured a 6km-deep rupture of the German front. According to the commander of the 12th SS-Panzer Division Hitlerjugend – Colonel Kurt Meyer – this advance had smashed the 89th Infantry Division, thus creating a yawning gap in the German front that remained undefended and unoccupied.

During the late morning of 8 August, Simonds' offensive began its second phase – the daytime break-in operation directed against the enemy's reserve defence line located between Bretteville and Saint-Sylvain. His two reserve armoured divisions moved south of Caen, ready to move up to the front to spearhead this assault. Erroneously believing this second German position to be a formidable one, Simonds had arranged that a second bombing strike should smash the enemy line shortly after noon. That morning 681 American B-17 Flying Fortresses droned their way south from England towards the Falaise plain, striking six German targets between 1226hrs and 1355hrs.

The late morning of 8 August thus witnessed a largely unavoidable lull in the Allied advance as Simonds' forces readied themselves for the second phase while waiting for the bombers to arrive. Unfortunately for the Allies, during this lull the enemy began recovering from the shock inflicted upon them by the successful night attack. Orchestrated by Meyer, the Germans now launched counter-measures against the Allied penetration.

MEYER DECIDES TO ATTACK

With Simonds' forces ready to strike south as soon as the approaching strategic bombers had crushed the reserve German defence line, Meyer and his subordinate Hans Waldmüller realised the situation was critical: if the Allied armour struck south, it would rupture the as yet still thinly manned German reserve position and charge south to capture Falaise. Simonds' forces might subsequently thrust south to link up with the north-westerly advance of American forces from Argentan, enveloping an entire German Army in what would become known as the Falaise Pocket. Meyer knew he had one choice – to throw whatever meagre forces he had in the vicinity into a desperate, probably suicidal, charge north into the Allied lines. Despite facing enormous odds, Meyer hoped that through the sacrifice of his forces, he might buy some precious time to enable other German units to move north and shore up the largely unmanned second defensive line.

At around noon, Meyer ordered his forces to initiate an immediate, improvised counter-strike against the Allied line at Saint-Aignan-de-Cramesnil. Unfortunately, Meyer had few forces available in the vicinity to initiate this audacious impromptu response. Operating under Waldmüller's command, this scratch force was a composite of several units. The infantry component comprised 500 Panzergrenadiers from 1st Battalion, 25th SS-Panzergrenadier Regiment. The stiffening element for this force came from 20 tanks (mostly Panzer IVs) from 12th SS-Panzer Regiment. The vanguard of the counter-strike, however, was formed from four or five (one source suggests seven) Tigers from Captain Michael Wittmann's 2nd Company, 101st Heavy SS-Panzer Battalion. Meyer was hoping that this desperate mission might achieve success through the lethal firepower and proven battlefield survivability of these few Tiger Is. Although the evidence concerning the composition of this Tiger force is contradictory, it is certain that most of the tanks were from Wittmann's own troop. However, a mass of Allied units faced this weak German strike force. In the La Jalousie–Saint-Aignan-de-Cramesnil sector the Allies deployed three armoured regiments and four infantry battalions, not to mention the spearhead of two armoured divisions assembling behind them. Although only a small proportion of these tanks could bring their fire to bear on Wittmann's Tigers, his armoured column was clearly about to charge north into an inferno of Allied tank, anti-tank and artillery defensive fire. At the moment when Meyer ordered Waldmüller to initiate this ad-hoc counter-attack, Wittmann was in his Tiger I, alongside three others, located in the Les Jardinets area, 600m east-south-east of Meyer at Gaumesnil. One of the major benefits to the Germans in their defensive role was their familiarity with the territory.

Carefully camouflaged, the drivers had hidden their Tigers behind a typical tall Normandy tree-hedge line; the row of low bushes running along the other side of this narrow dirt lane added to this concealment. At around 1205hrs, Wittmann heard that Meyer had urgently summoned him to a briefing at nearby Cintheaux. Wittmann's driver steered Tiger 007 west down country lanes for a few minutes to arrive at the village. Jumping down from his turret, Wittmann found Meyer and Waldmüller settling the final details of the counter-strike, outlining the part Wittman's Tigers would play in the action.

According to Meyer, a dramatic event then occurred that transformed the timeframe for the intended attack. The three SS officers observed a solitary Allied heavy bomber fly over Cintheaux, sending out coloured flares. This was obviously an Allied pathfinder designating the aim points for an impending strategic bombing strike. Given what Meyer already knew about Allied pathfinder tactics from the earlier Normandy battles, he concluded that the heavy bombers were already less than ten

ABOVE One of only six surviving Tiger Is, this tank, abandoned in the last days of the battle of Normandy, is displayed at the entrance of the French city of Vimoutiers. The colour scheme was applied by a local volunteer and does not accurately reflect the paint of the Tigers in use in Normandy. (TCY)

minutes away from their current location. It was obvious to all that the Allied armour assembling behind the current front line had not commenced its assault south because they were waiting for heavy bombers to unleash a rain of destruction on the German forces below.

This realisation merely confirmed Meyer's recent decision to strike north. He once again ordered his men to launch the attack – but now to do so immediately. Despite the appalling odds facing his mission, Wittmann climbed into his tank and headed east back to his company's positions in Les Jardinets. Within minutes, his force of four Tiger Is had shaken off their camouflage and had begun to advance north across the open fields towards the Allied lines. A few hundred metres further east, a fifth Tiger I (and possibly two others) also began to rumble north. Over the ensuing hour Wittmann and his Tigers bravely attempted to wrest an improbable victory from the jaws of defeat.

ADVANCE TO SAINT-AIGNAN-DE-CRAMESNIL

One of the Allied units deployed in the La Jalousie–Saint-Aignan-de-Cramesnil sector would play a key role in defeating Wittmann's riposte. The 1st Northamptonshire Yeomanry was a Sherman-equipped armoured regiment in the independent British 33rd Armoured Brigade. Led by Lieutenant-Colonel D. Forster, the regiment fielded three squadrons, 'A', 'B' and 'C', plus the regimental headquarters. Most of the regiment's tanks bore distinctive names: 'A' Squadron's were named after Soviet towns, 'B' Squadron's after American states and 'C' Squadron's after Northamptonshire villages. The regiment's authorised strength was 59 Shermans, including 12 Fireflies.

As part of the 'left British column', which also comprised the infantry of the 1st Black Watch, the 1st Northamptonshire Yeomanry had advanced 6km to Saint-Aignan-de-Cramesnil during the early hours of 8 August. Having assembled in a 50-vehicle-deep by four-vehicle-wide formation, the column's advance commenced around midnight. At 0130hrs, No. 2 Troop, 'A' Squadron, which had detached from the rest of the column, stumbled upon four well-camouflaged enemy self-propelled guns (SPGs) and a fierce action ensued that demonstrated the Firefly's killing power. Four rounds hit Lieutenant Jones's lead Sherman No. 5 'Brest-Litovsk'; one tore off the external blanket bin, another penetrated to lodge in the vehicle's transmission gear, the third gouged a groove out of the frontal armour before bouncing away and the fourth smashed through to the engine. With the tank on fire, Sergeant Burnett's IC (Hybrid) Firefly No. 8 'Balaclava' engaged the three SPGs. The Firefly dispatched two of them with three rounds, but not before the third SPG had knocked out the troop's two remaining standard Shermans. While the sole survivor – Burnett's Firefly – caught up with the column, the three dismounted crews spent the rest of the night pinned to the ground by machine gun fire. Lieutenant Jones had previously written to his mother telling her not to worry, but during this ordeal he said out loud: 'for goodness' sake, mother, start worrying now!'

Subsequently, the column rumbled on until, at 0325hrs, it reached a hedge situated north of its objective, Saint-Aignan-de-Cramesnil. Accompanied by artillery fire, two Sherman squadrons pushed through gaps in the hedge and engaged the enemy, while the now dismounted infantry stormed the village. After an hour-long battle, the column captured Saint-Aignan-de-Cramesnil and began establishing firm defensive positions around it. As part of this process, 'A' Squadron assumed defensive positions in the orchards located south-west of Saint-Aignan-de-Cramesnil at Delle de la Roque. At this stage the regiment fielded 54 operational Shermans, including the 12 Fireflies, having lost four standard Shermans during the night advance; a fifth tank – Sergeant Duff's No. 56 'Lamport' – ended up with the 144th Regiment Royal Armoured Corps at Cramesnil.

WITTMANN'S DEATH CHARGE

At around 1220hrs, Wittmann's troop of four Tiger Is began to advance north from Les Jardinets. That day, Wittmann was not in his usual vehicle, No. 205, which was being repaired. Instead, he was using battalion commander Heinz von Westernhagen's command Tiger, No. 007. Apart from Tiger 007, it is probable

SHERMAN VC FIREFLY SPECIFICATIONS

GENERAL
Production run: January 1944–May 1945 (17 months)
Vehicles produced: Between 2,139 and 2,239
Combat weight: 35.36 tonnes
Crew: Four (commander, gunner, loader/radio operator, driver)

DIMENSIONS
Overall length: 7.82m
Hull length: 6.5m
Width: 2.67m
Height: 2.74m

ARMOUR (THICKNESS AT DEGREES FROM VERTICAL)
Hull front: 51mm at 45–90°
Hull sides: 38mm at 90°
Hull rear: 38mm at 70–90°
Hull roof: 25mm at 0°
Turret front: 38–76mm at 85–90°
Turret sides: 51mm at 85°
Turret rear: 64mm at 90°
Turret roof: 25mm at 0°

ARMAMENT
Main gun: 1 x 17 pounder (76.2mm, 3in.) ROQF Mk IV or Mk VII

Secondary: 1 x 0.3-calibre co-axial M1919A4 machine gun
Main gun rate of fire: 10rpm

AMMUNITION STOWAGE
Main: 77 rounds (typically 50 per cent APCBC/APDS, 50 per cent HE)
Secondary: 5,000 rounds

COMMUNICATIONS
No. 19 Set transmitter/receiver

MOTIVE POWER
Engine: Chrysler Multibank A57 30-cylinder petrol engine
Power: 443 metric bhp at 2,850rpm
Fuel capacity: 604 litres
Power-to-weight ratio: 12.9hp/tonne

PERFORMANCE
Ground pressure: 0.92kg/cm^2
Maximum road speed: 36kph
Maximum cross-country speed: 17kph
Operational range (road): 201km
Operational range (cross-country): 145km
Fuel consumption (road): 3 litres/km
Fuel consumption (cross-country): 4.2 litres/km

that the column comprised Ihrion's Tiger 314, Dollinger's 008, and Blase's tank; the exact composition of and the order within this formation remains unclear. The tanks rumbled north-north-west, one behind the other, on an axis parallel to, but around 150m east of, the main Caen–Falaise road. That Wittmann's column was in line ahead suggests that he expected enemy fire from the north around Hill 122. This deployment, however, left the Tigers' more vulnerable flanks exposed to fire from the north-east – from the orchards of Delle de la Roque, south-west of Saint-Aignan-de-Cramesnil. Wittmann remained unaware that 'A' Squadron, 1st Northamptonshire Yeomanry, was positioned in these woods.

Both Meyer and his medical officer, Captain Wolfgang Rabe, observed the initial stages of Wittmann's advance. The Tigers rumbled north through a hail of Allied defensive artillery fire. The tanks only stopped periodically in shallow gullies to fire at long range towards the north-west to engage tanks of the Canadian Sherbrooke Fusiliers Regiment located west of the main road. At a range of 1,800m, the Tigers knocked out several Shermans; at this distance, only the Canadian regiment's few Fireflies stood a slim chance of destroying the Tigers. Having observed the first phase of Wittmann's charge from the northern fringes of Cintheaux, Meyer's attention was now diverted to the northern horizon. Suddenly, the sky to the north began to darken as what seemed to be an endless stream of Allied bombers droned south towards the village. Meyer and the grenadiers dug in around him were rendered speechless by this display of vast Allied air power.

'What an honour,' remarked one. 'Churchill is sending a bomber for each of us!' The Panzergrenadiers raced across the open fields located north of the village to escape the impending onslaught. Just in time, they witnessed the bombers pass over them and begin dropping their bombs onto the village behind them.

Deployed behind the cover provided by the tall tree-lined hedge that marked the southern border of the Delle de la Roque orchards, the four tanks of 'A' Squadron's No. 3 Troop held the westernmost part of the 1st Northamptonshire Yeomanry defensive position. Led by Lieutenant James in tank No. 9, a standard Sherman, the troop also deployed two more standard Shermans – Sergeant Eley's No. 10 'Vladivostock' and Corporal Hillaby's No. 11. The troop's last and most potent asset was Sergeant Gordon's VC Firefly No. 12, according to some sources, named 'Velikye Luki'. Straining to see through the gaps in the tree-hedge, the troop's observers peered south across the open ground towards the Les Jardinets area.

Three Tigers were heading north-north-east in line ahead, on an axis just east of the main road. The range was 1,200m. Gordon reported this sighting via radio to Captain Boardman, the squadron second-in-command. Boardman claims that he then ordered Gordon to hold his fire until he could get there to direct the action. Why the captain did not instruct the squadron's three other Fireflies to reinforce Gordon's tank remains a mystery. Moving west through the orchard in his Sherman Mk I, No. 18 'Omsk', Boardman soon arrived at the No. 3 Troop position. By this time rather random German artillery and mortar fire was landing in the general area of the orchard.

The British observed the advancing Tigers for a couple of minutes. The enemy seemed to be unaware of the British tanks' presence, as the Tigers still remained deployed in line ahead on a bearing north-north-east, which exposed their relatively less formidable side armour to No. 3 Troop's tanks. Convinced that they were undetected, the troop calmly allowed the Tigers to advance until the range had closed to 800m, by which time the tanks had neared the vicinity of the isolated red-roofed building adjacent to the main road. At this range, Gordon's Firefly stood a good chance of penetrating the Tiger's side armour, although the other Shermans still stood virtually no chance at all. The time was now 1239hrs. According to Gordon's gunner, Trooper Joe Ekins, the sergeant now told the other Shermans to stay under cover while he courageously attempted to deal with the Tigers.

Gordon ordered his driver to move the Firefly forward a few metres to a position just in front of (that is, south of) the orchard's southern edge to obtain a better field of fire. Gordon remained with his head poking through the open commander's cupola during the ensuing brief action. Gordon selected his target – the rear Tiger of the three, as was normal practice; such a tactic hoped to exploit the fact that the leading tanks might not even know that their rear colleague had been hit. The reason why British accounts only mention three Tigers, when Wittmann's column may well have had four or five, remains unclear. Some of the Tigers may have dropped out of the column and were heading north-north-east on an axis to the left (west) of the others and may even have been obscured from view. The time was now 1240hrs. Looking through his sight, gunner Ekins was now very frightened because he believed that there 'was no way' a solitary Firefly could take on three Tigers and survive. With 'but one thought in my mind – to get the bastard before he gets you' – Ekins aimed the gun and fired two armour-piercing rounds at the rear Tiger. Despite having only fired six 17-pounder rounds before, Ekins nevertheless had 'a knack' at gunnery, and both rounds seemingly hit the target; within seconds, the Tiger was burning. Other Allied tanks, however, were also engaging the Tigers at this time. The Canadian Fireflies of the Sherbrooke Fusiliers were firing from their positions west of the main road at a range of 1,100m. Similarly, the Fireflies of 144th Regiment Royal Armoured Corps, located on Hill 122, were engaging Tigers at a range of 1,300m. While it is not impossible that this longer-range fire hit and penetrated the Tiger at precisely the same as Ekins engaged it, the most likely explanation is that Ekins's rounds penetrated the tank and caused it to burn.

As soon as the Firefly had fired its second shot, Gordon followed doctrine by ordering the driver to reverse back into the cover of the orchard. As they did so, the second Tiger traversed its gun right towards the Firefly. Looking through his sight, Ekins recalled that the Tiger's 8.8cm gun 'looked as big as a battleship' as it swung to face him. The Tiger fired a round at the Firefly as it began to reverse and then a further two

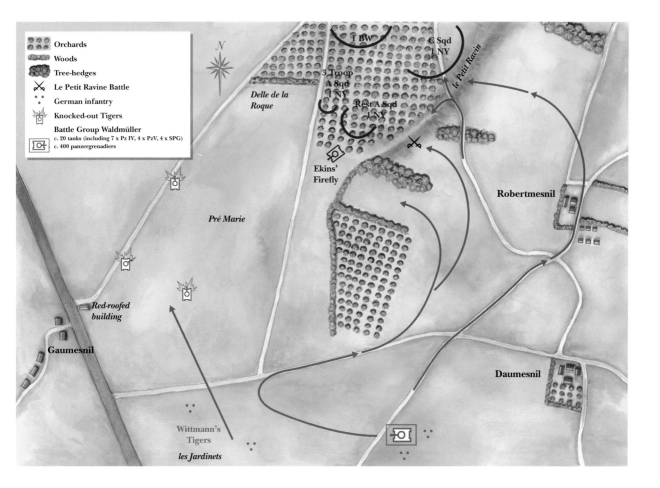

Orchards

Woods

Tree-hedges

✕ **Le Petit Ravine Battle**

∴ **German infantry**

⊡ **Knocked-out Tigers**

Battle Group Waldmüller
⊡ c. 20 tanks (including 7 x Pz IV, 4 x PzV, 4 x SPG)
   c. 400 panzergrenadiers

N

1 BW

C Sqd
1 NY

3 Troop
A Sqd
1 NY

*Delle de la
Roque*

Rest A Sqd
1 NY

le Petit Ravin

Ekins'
Firefly

Robertmesnil

Pré Marie

*Red-roofed
building*

Gaumesnil

Daumesnil

Wittmann's
Tigers

les Jardinets

rounds as the tank entered the concealment of the apple orchard. As the third round passed close by the tank, the flap of the open commander's cupola came crashing down onto Gordon's head, knocking him half-senseless. It is not clear if this was caused by the flap knocking into a tree-branch or because the Tiger's round had actually hit it a glancing blow. The dazed Gordon clambered uneasily down from his tank and was immediately wounded by shrapnel, as the German artillery and mortar fire moved closer to the tree-hedge. All Ekins knew of this incident was that suddenly there was no commander in his tank!

Next, the commander of No. 3 Troop – Lieutenant James – bravely jumped out of his tank and raced across to take command of the Firefly. James ordered the driver to move the tank to a new firing position. The tank reached this new position just before

1247hrs, according to the war diary entry. Moving out from cover, James now ordered Ekins to engage the second Tiger – the one that had fired at the Firefly.

At 1247hrs Ekins fired one shot at the second tank, which hit it, causing it to explode in a ball of flame. As the tank Wittmann was believed to be in – command Tiger 007 – was found with its turret blown off, and German eyewitness reports recorded only one exploding tank, it seems likely that this was the precise moment at which Wittmann's prolific career was terminated. Next, Gordon's driver again reversed the Firefly back into the cover of the orchard. This success just left intact the lead Tiger and the mystery fourth and fifth Tigers, which the British accounts of the battle seem to have missed altogether.

At this juncture, it is claimed that some standard Shermans advanced south out of the woods to engage the remaining Tigers at such close range that they stood some chance of damaging them, but this inadvertently hampered the fire delivered by Ekins's Firefly, which had re-emerged out of the orchard. The ensuing hail of

75mm Sherman fire apparently fell upon the lead enemy tank. While these rounds failed to penetrate the tank's thick armour, the hail of fire caused the driver to veer off erratically to the west, seemingly out of control. According to the regimental historian this tank 'was in a panic, milling around wondering how he could escape'. Captain Boardman then claims that he engaged the veering Tiger with a 75mm AP round that caused it to stop. Ekins, however, states that 'it was still moving when I hit him'. Ekins fired two shots that caused the Tiger to burst into flames; from this inferno none of the crew escaped. The time was 1252hrs. The final successes of this action also went to the Firefly. Just eight minutes later, at 1300hrs, Sergeant Finney's Firefly – tank No. 4 'Orenburg' – spotted two Panzer IVs moving to the west side of the main road at the prodigious range of 1,645m. In a brilliant piece of shooting, gunner Trooper Crittenden fired two shots and 'brewed up' both Panzers.

Wittmann's death charge had been a remarkable action. As Ekins well knew, the chances of a solitary Firefly surviving a clash with not just one but at least three (and possibly five) Tigers were extraordinarily slim. Yet in the space of just 12 minutes, Gordon's Firefly had dispatched three Tiger Is with just five rounds. In return, not only did the Tigers fail to knock out a single enemy tank, but it seems as if not one of the rounds they fired hit its target. This astounding feat was one of the finest tank-versus-tank engagements seen during the entire North-West Europe campaign. It seems odd, therefore, that none of his peers congratulated Ekins for this feat, even though the gunner maintains that he never expected it anyway. The accomplishment also deserves more recognition than the laconic note penned in the regiment's war diary: 'Three Tigers in twelve minutes is not bad business.' The combination of three factors – the regiment's use of terrain to ambush the Tigers, Gordon's 'knack' at gunnery and the Firefly's awesome gun had turned the dispatching of one of Germany's finest aces from an almost impossible task into something 'rather like Practice No. 5 on the ranges at Linney Head'.

DEATH OF TIGER 007

On receipt of orders from tank commander Lieutenant James, the Firefly's gunner – Trooper Ekins – begins to lay the 17-pounder gun on the target, Wittman's Tiger.

Once the Tiger is in the cross hairs, Ekins calculates the range – around 800m – and the loader places an AP round in the breech. As James counts down '3–2–1–fire!', the crew place their hands over their ears, open their mouths and close their eyes – all in preparation for the violent back blast.

The gun is fired and within a second the round has smashed into the Tiger. For a few seconds, however, this fact remains unknown to the Firefly crew, as they are momentarily blinded by the flash and back blast.

The blast having subsided, Ekins can see smoke rising from the Tiger through his gunsight. A few more seconds pass and then Ekins sees the Tiger erupt as its ammunition explodes.

August 1944, 1247hrs

+50 seconds

What then was Wittmann's fate? Dr Rabe had observed this short but bloody battle from the western side of the main road. He recalled that the four or five Tigers involved had come under enemy fire and that several had gone up in flames. He attempted to get closer to see if any of the crews had survived, but could not because of enemy fire. After waiting for two hours, not a single crewmember had emerged from the battlefield, and so Rabe withdrew, assuming that all had been killed. Wittmann was officially listed as missing in action.

The up-gunned Sherman had proved its ability to vanquish the most-feared German heavy tank, the Tiger. From this duel at Saint-Aignan-de-Cramesnil – probably the last great clash of Firefly versus Tiger – the Firefly emerged triumphant.

THE CONTROVERSIES OF WAR

Many events that occur in battle are subject to different interpretations thanks to the inherent 'fog of war' and the differing deductions made from the available evidence. Nowhere is this more evident than when famous wartime figures meet their death in battle. Given Wittmann's legendary status it should not surprise us that controversy has raged over which unit and which weapon actually destroyed his tank. For many years after 1945, however, no one on the Allied side even realised that Wittmann had been killed during the 8 August battle, which thus remained just one of hundreds of otherwise unremarkable, half-remembered, wartime actions.

Once it became known on the Allied side that Wittmann had been a victim of this battle, the leading interpretation was that his demise had been caused by a high-explosive rocket fired from an RAF Hawker Typhoon aircraft. This interpretation was based largely on the circumstantial evidence that an unexploded rocket was found nearby, and on the specious logic that such a devastating explosion could only have been caused by such a weapon. As the doubts about this explanation mounted, a number of British and Canadian armoured regiments deployed near Gaumesnil that afternoon claimed the distinction of dispatching Wittmann; of these – the claim of the 1st Northamptonshire Yeomanry – stood out as being the one best supported by convincing contemporary evidence.

As we have seen, the 1st Northamptonshire Yeomanry war diary recorded three Tigers being destroyed at the exact time when, and in the general location where, Wittmann's Tiger was destroyed. Furthermore, the unit's account of the battle was produced within a few weeks of the battle when no one in the regiment realised the significance of what they were describing. No other interpretation of how Wittmann came to be killed can remotely compete with the wealth of unequivocal, impartial, contemporary evidence that supports the claim of the 1st Northamptonshire Yeomanry. In all probability, it was a woefully inexperienced Firefly gunner, Joe Ekins, who dispatched the veteran Tiger ace that afternoon.

AFTERMATH AND ANALYSIS

Thanks to the Firefly, Wittmann's desperate charge north into the Allied lines near Saint-Aignan-de-Cramesnil was repulsed and his Tigers either destroyed or abandoned. Allied Firefly and standard Sherman fire, augmented by anti-tank and artillery fire, also eventually forced back the other element of Meyer's counter-attack force in the battle that raged in and around Le Petit Ravin. Meyer's audacious attempt to block the impending Allied armoured onslaught in the second phase of Operation *Totalize* had failed. The Allied armour commenced its attack south on schedule, after the bombing ended at 1355hrs. In these circumstances, the rest of 8 August 1944 ought to have been precisely the disastrous 'black day' for the Germans that the Canadians had hoped it would be. Thanks to the Firefly, Meyer's fear that the Allies would successfully race south to occupy Falaise that day ought to have been realised.

But the spectre of disaster did not materialise. The two armoured divisions of Lieutenant-General Guy

Simonds' II Canadian Corps – both as yet unblooded in the horrors of combat in Normandy – advanced cautiously that afternoon, fearful of the long-range killing power of the handful of Tigers, Panthers, Panzer IVs and 75mm anti-tank guns still available to the meagre defending German forces. The lethal firepower of these assets in such open terrain was soon demonstrated when the advancing Polish armour lost 40 tanks to German fire in the space of just 15 minutes. The defenders used the intervening time well to rush reinforcements to their severely depleted front on the Falaise plain. In fact it would take Simonds' forces another week to secure the high ground that dominated Falaise. Thus it was a combination of other factors, plus some good fortune – rather than the sacrifice of Wittmann's Tigers – that enabled the German forces

OPPOSITE German Tiger and Panther tanks stand in a churned-up field where they were knocked out by American fire north of the village of Saint-Lô after an Allied breakout from invasion beachhead areas. (Photo by Frank Scherschel/The LIFE Picture Collection/Getty Images)

to escape the debacle envisaged by Meyer that lunchtime on 8 August.

This lucky escape, however, did not alter the fate that was about to consume the German Western Army in Normandy – it merely postponed it by a few days. By mid-August, Simonds' forces had pushed forward beyond Falaise to link up with the Americans near Trun to close the Falaise pocket. Undoubtedly the Allies had won a great victory.

The German Seventh Army and the greater part of the Fifth Army had been destroyed with the loss of thousands of German soldiers. Although many German troops escaped with some armour and guns towards the Seine, no significant German forces remained to face the Allies after the loss of Normandy. In the ensuing weeks the Allies rampaged forward all the way to the borders of the Reich. There seemed a remote and fleeting opportunity that the war might be won during 1944. However, Allied logistical exhaustion and rapid German recovery ensured that the Allies would have to grind their way through

Germany in a series of bitter battles that would rage into early 1945.

So what did the summer 1944 battle for Normandy – and particularly Wittmann's 8 August 1944 death charge – prove about the epic clash of Firefly versus Tiger I? The Normandy campaign aptly displayed the weaknesses of the expedient Firefly design. With armour identical to the standard Sherman, the Firefly remained vulnerable to enemy tank and anti-tank fire. It took great bravery from Sherman crews to go into action against the latest generation of powerfully gunned Panzers, especially given the Sherman's legendary tendency to burn when hit. Yet equally, the campaign – and the 8 August action in particular – showed that, in the Firefly, Allied armoured units had finally got the 'Panzer killer' they required. These battles showed that the Firefly could take on and defeat all of Germany's latest tanks, including the much-feared Tiger. The battle that raged near Saint-Aignan-de-Cramesnil on 8 August represented the Firefly's finest hour.

In August 1944, therefore, the Firefly and the Tiger were the dominant tanks on the battlefields of North-West Europe.

Yet this dominance was short-lived. While from the perspective of Normandy the Firefly emerged as the victor and the Tiger I the vanquished, both at this time shared a growing sense of approaching demise. With its production ended and the tank now outclassed by the Tiger II, the Tiger I would appear on the battlefield in decreasing numbers until the end of the war. So too the Firefly; with production ended in May 1945, it was soon replaced by specifically designed medium tanks like the Comet that were better armed and armoured. The Firefly was, after all, just an expedient – albeit an economical, effective and well-timed one. Without its lethal firepower, it is possible that the German Western

OPPOSITE British Engineers fill the wreck of a German Tiger I tank with landmines in order to destroy it, after the Allies recaptured the village of Villers-Bocage in Normandy, site of Michael Wittmann's famous victory in June 1944. (Photo by Leonard McCombe/Picture Post/Getty Images)

Army may have been able to maintain a coherent front in Normandy for much longer than they did. Sherman Fireflies were indeed, as reports at the time suggested, 'battle-deciding weapons … every one of which … will help materially to shorten the war'.

TIGER II AND IS-2 – EAST PRUSSIA, 1945

THE STRATEGIC SITUATION

Following the encirclement and destruction of Axis forces at Stalingrad, Soviet forces spent the next two years conducting an inexorable strategic advance that recaptured all of their lost territory, and continued beyond their pre-1939 borders into Eastern Europe. Supplied with substantial Lend-Lease logistical assets, especially badly needed transport, the Red Army was frequently able to concentrate overwhelming numbers of men and machines at sectors and times of its choosing. Its effective use of *maskiróvka* ('deceptive measures') and a growing level of operational experience meant that the Germans were often forced to react to an unexpected or unfavourable situation. With Soviet forces seemingly in strength everywhere along the front, German command and control could not consistently or effectively prioritise and address threats and their combat formations were frequently forced to withdraw or risk destruction.

During the summer of 1944 this scenario occurred on a grand scale where the Soviet *Bagration* offensive virtually annihilated Army Group Centre. Hitler's belief that the Red Army would try to use its recent acquisition of western Ukraine as a springboard from which to attack into Romania, Hungary and southern Poland prompted him to reposition much of his armour south to contest such a move. Instead, the Soviets launched a devastating offensive further north that drove a massive wedge into Belorussia. Unable to regain their strategic balance following Stalingrad and Kursk, the German Eastern Army (*Ostheer*) traded space for time, while they tried to re-establish adequate defences and solidify the front. To compound their difficulties, Hitler continually interfered with the war's conduct and the actions of experienced commanders on the spot.

Hitler, in the face of strong advice and intelligence that Berlin would be the target, believed that the Soviets' next great offensive would be against his East Prussian and Hungarian flanks. As a consequence, Marshal Georgy Zhukov's 1st Belorussian and Marshal Ivan Konev's 1st Ukrainian fronts were able to capitalise on their achievement against far fewer opposing forces than would have otherwise been possible.

In mid-January 1945 the pair were able to quickly overrun Poland and establish several small bridgeheads across the Oder River from near Stargard to south of Breslau. To help stem the flood of Soviet forces moving across Poland, an emergency army group, Vistula (*Weichsel*), was created to bolster the much-weakened Army Group Centre; instead of an experienced commander, though, Hitler assigned Reich Leader-SS Heinrich Himmler to the task. German reinforcements were soon moving through Stettin, among them III (Germanic) SS-Panzer Corps assigned to act as the linchpin of Himmler's command.

Instead of deploying his forces across the most direct route to Berlin, however, Himmler arrayed the various subordinated replacement and recently separated units in parallel to the Baltic coast in an effort to defend the whole of Pomerania. Zhukov, confident in his superiority of 3:1 in infantry and 5:1 in armour and artillery, simply ignored what he believed to be a 'phantom' force, and continued to focus westward towards the German capital.

As Konev approached the Oder, Zhukov's forces encircled the city of Posen, a major transport hub, on 24 January. As with other Hitler-designated 'fortresses' (*Festungen*) the city was expected to be held at all costs, in part to draw large numbers of the enemy into static, attritional battles that favoured the defender.

On 27 January lead Soviet armoured units crossed the Draga River at Neuwedell, appearing to herald the collapse of the Pomeranian front. As the front line receded towards towns like Arnswalde, Reetz and Stettin, residents tried to help the passing refugees, with available trucks and trains removing any civilians they could. When the Red Army cut the direct routes between Deutsch Krone and Stargard on 4 February, such operations through the Arnswalde area largely ceased.

GERMAN PREPARATIONS

Despite all the military setbacks Germany had suffered over the last two years and their increasingly short supply lines, the effective resource management by Albert Speer and others permitted a rapid reaction to the faltering Eastern Front. Between 20 January and 12 February the German Navy and hundreds of transport ships, as part of Operation *Hannibal*, had evacuated some 374,000 refugees and thousands of wounded soldiers from sectors cornered against the Baltic Sea and threatened with annihilation. Salvaged combat formations were redirected to Pomerania's defence. In early February, Army Chief of Staff Heinz Guderian looked for a way to pinch off the Soviet spearheads that had advanced to the Oder River before follow-on forces could strengthen the positions. With his hopes for a two-pronged operation from the Küstrin and Stargard directions quashed by Hitler's refusal to permit the reallocation of troops from areas such as Italy, Norway and the Balkans, Guderian settled on assembling the remaining northern force to attempt a more modest operation.

OPERATION *SOLSTICE*

Despite the chaotic military situation facing Germany, by early February Guderian was able to assemble a surprisingly large force consisting of three and a half divisions from Third Panzer Army and two re-formed Panzer divisions. SS-Senior Group Leader Felix Steiner, an experienced battlefield commander, was put in charge of the hastily organised, corps-sized Eleventh SS-Panzer Army.

On 8 February, Stalin abruptly cancelled the impending offensive against Berlin, and instead

OPPOSITE A Tiger II of the 505th Heavy Panzer Battalion. The battalion was reequipped with Tiger IIs after the Soviet offensive during Operation *Bagration*. (Panzerfoto)

ordered that Marshal Konstantin Rokossovsky's 2nd Belorussian Front first clear Pomerania. Although logistical problems and encircled German 'fortresses' presented temporary obstacles, Stalin was motivated by political considerations as well. The Western Allies had been held up west of the Rur River and the *Westwall* ('Siegfried Line' to the Allies) for the last six weeks by Germany's counter-offensive in the Ardennes, and Eisenhower's forces were only now renewing their eastward drive into Germany and would not come near those territories that Stalin coveted for some time.

After eight days of stockpiling food, fuel and ammunition for Guderian's counter-attack, by 10 February less than half of the estimated requirements had been obtained. With the attack scheduled for the 22nd, Third Panzer Army was not likely to arrive in time to participate. Eleventh SS-Panzer Army – with its subordinate XXXIX Panzer, III (Germanic) SS-Panzer and X SS-Corps – would be all that was available for carrying out the overly optimistic mission of fighting through the Soviet Sixty-First Army and

advancing to the Küstrin–Landsberg area to cut off Second Guards Tank Army spearheads.

Hitler and Himmler were reluctant to commence such an operation until sufficient supplies had been gathered, but Guderian managed to secure approval for an amended start date. To gain a measure of operational control, and provide the greatest chance for success, he had his young (but skilled and experienced) protégé, General-Lieutenant Walther Wenck allocated to command the operation, which was codenamed *Hussar Ride* (*Husarenritt*), later changed to *Solstice* (*Sonnenwende*).

POMERANIA

To bolster German defences along the Oder River and east of Stargard against Semyon Bogdanov's Second Guards Tank Army as it rampaged across southern Pomerania, 503rd Heavy SS-Panzer Battalion was

OPPOSITE A Tiger II, with spare track links hung on the turret waits in the cover of a forest. (Jim Laurier, © Osprey Publishing)

activated at its bases around Berlin on 25 January 1945. With so many trained crews having been sent to the 501st and 502nd Heavy SS-Panzer battalions over the previous weeks, the formation was hard-pressed to provide occupants for its Tiger IIs when word suddenly came that it was to depart for the front line between Friedberg and Schneidemühl. As additional crews were procured, the battalion's new commander, SS-Storm Command Leader Friedrich 'Fritz' Herzig, implemented final preparations to enter the fray the following day.

On 26 January, after months of intensive training, and having received their final shipment of 13 Tiger IIs from the Army Ordnance Department at Kassel, 503rd Heavy SS-Panzer Battalion entrained for the short ride to the Eastern Front with 39 vehicles. Although doctrine dictated the desirability of allocating such units as a whole to maximise their battlefield potency, the number of threatened sectors along the porous Pomeranian front line pressured German commanders into parcelling elements off as circumstances dictated. As the battalion passed through Berlin its 2nd Company's 1st Platoon was diverted to the Küstrin bridgehead defence as the remainder of Herzig's command continued on.

Over the next few days the Germans continued to rush forces to Pomerania to establish some sense of order. The unexpectedly rapid Soviet advance across Poland had overwhelmed the defenders.

As 503rd Heavy SS-Panzer Battalion entered their assigned sector east of Stargard on the 28th, the amorphous front line added to the confusion as the formation's platoons spread out to strengthen the area's defences. After being allocated to Major-General Oskar Munzel's command, Herzig accompanied his Headquarters Company and a dozen Tiger IIs as they detrained at the Pomeranian district capital of Arnswalde. The remainder of 503rd Heavy SS-Panzer Battalion pushed ahead along the northern edge of the Warthe River. Six vehicles from 3rd Panzer Company were shuttled further south towards Friedberg, while three more continued on to Schneidemühl. The remaining Tiger IIs made up Lieutenant Max Lippert's 1st Panzer Company east of Reetz.

With Red Army forces having crossed the Warthe River at Landsberg and Driesen, 3rd Panzer Company battle group continued south to help contain the bridgehead. With Soviet armoured patrols expected in the area, the Tiger IIs were ordered by the local commander, Major-General Kurt Hauschulz, to detrain prematurely west of Friedberg and continue towards their destination under their own power. As head of Sixteenth Army's NCO school at Stargard, Hauschulz had recently rushed some 800 of his cadets to counter Soviet penetrations between Arnswalde and Schneidemühl. Unfortunately for the Germans, the 'Pomeranian Wall' had little more than scattered, low-quality *Volkssturm* militia units with which to defend it, the majority of fortress units having been sent to defend the *Westwall* against British and American encroachment. Not surprisingly, Soviet advances from the south and east quickly made these defences untenable. The six-vehicle battle group would soon fight off Soviet armoured probes and anti-tank positions around Heidekavel, and disrupt the supplies moving through the area en route to the Second Guards Tank Army at the Oder. Additional enemy units, however, soon forced the overextended Germans to pull back.

With the remainder of 503rd Heavy SS-Panzer Battalion deployed east of Arnswalde, Himmler ordered a portion of the unit to organise blocking positions north-east of Reetz against expected Soviet probes. A half-dozen Tiger IIs from Lippert's 1st Panzer Company set off against sporadic resistance towards the Driesen bridgehead with one of the anti-aircraft platoon's three 4 × 20mm Flakpanzer IV *Wirbelwind* self-propelled guns. Along with 365 grounded paratroopers from Fallschirmjäger Regiment zbV Schlacht, the battle group merged with elements of a reconnaissance and an anti-aircraft battalion at Neuwedell before heading south to win back the recently Soviet-occupied town of Regenthin.

Further east, 2nd Company's 3rd Platoon deployed three of its four Tiger IIs and the battalion's anti-aircraft platoon at Schneidemühl. With one Tiger II soon succumbing to mechanical problems, the remaining vehicles moved to the eastern edge of town at Bromberger. Soviet artillery steadily shelled the surrounding area from the nearby Schneidemühl forest as the two Tiger IIs established hull-down positions behind a protective railway embankment. Leaving the immobilised tank to fight in the forthcoming siege of Schneidemühl, what was left of 2nd Company's 3rd Platoon was ordered to return to the west to help defend Küstrin. With a small contingent of protective infantry hitching a ride, the group avoided Soviet patrols as they made their way past Friedberg and Landsberg to reach their 180km-distant destination on 30 January.

On 29 January Major-General Hans Voigt was reallocated from his duties as commandant of the 'Pomeranian Wall' fortresses to that of Fortress Arnswalde to organise a motley collection of nearby units. Along with 400 local cadets, Battalion Enge and replacement staff for Artillery Regiment zV (*zur Vergeltung* – 'for retribution'), recently freed from their V-2 rocket-launching duties, provided 400 and 800 fighters respectively for Arnswalde's defence. Possessing little more than small arms and a few machine guns, these groups were bolstered by two *Volkssturm* battalions, a *Landesschütze* (Territorial Defence) battalion and a detachment from 83rd Light Anti-Aircraft Battalion that provided two anti-aircraft batteries. With only a handful of 81mm mortars and no artillery, Voigt's command relied on handheld anti-tank weapons and machine guns to provide support, as the former V-2 staff members established security positions at Hohenwalde, Klücken, Kürtow and Zühlsdorf. To the east the 1st Panzer Company battle group recaptured Neuwedell, but the effort resulted in high losses among the *Fallschirmjäger*. With a mix of NCO cadets, *Volkssturm*, emergency and other units under 402nd Division Staff zbV providing support, Lippert set out the following day with his available paratroopers and four Tiger IIs towards Regenthin. Numerous Soviet anti-tank guns and infantry, however, forced the Germans to withdraw to avoid being cut off.

TIGER II 'PORSCHE TURM', 503RD HEAVY PANZER BATTALION, RUSSIA, WINTER 1944-45

The 503rd Heavy Panzer Battalion picked up some of the last Tiger IIs from the Henschel factory on 31 March 1945. Tiger IIs of the 503rd Heavy Panzer Battalion fought in France against the Allied forces in the summer of 1944. Their Tigers were painted in a base dark yellow, with dark green and dark red stripes and patches added for camouflage. When pulled out of France on 9 September 1944 for redeployment to the Eastern Front, the 503rd was only able to save two of its 'Porsche' turreted Tigers, one of which was Tiger No. 314. This tank subsequently served in Russia during the winter of 1944–5. The hangers for spare track links were retro-fitted. In winter the troops were issued with water-based whitewash which they used to cover the tank in snowy conditions. This whitewash could be cleaned off when conditions changes. The original tactical signs and markings were not covered.

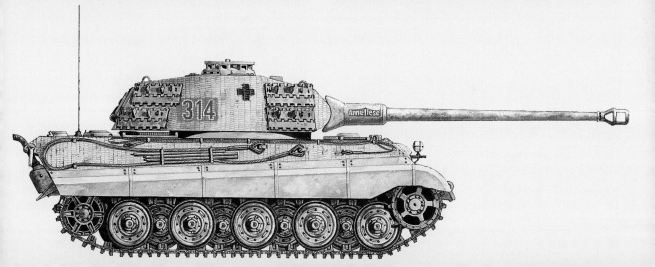

SOVIET DEPLOYMENT

Since pushing west from their bridgehead along the Vistula River on 16 January, Colonel Boris Eremeev's 11th Guards Heavy Tank Brigade made excellent progress as it punched through German defences ahead of the now ubiquitous T-34/85 medium tanks and mechanised forces. Increasingly urbanised combat and lengthening supply lines took a mechanical toll on the IS-2s, which were routinely doubling their design-estimated battlefield lives. Second Guards Tank Army, for example, lost 52 per cent of their tanks and self-propelled guns to enemy armour and artillery, and 43 per cent to German handheld anti-tank weapons such as the Panzerfaust and Panzerschreck in the first 24 days of the Vistula–Oder offensive. As a defence against the latter weapons, Soviet tankers began to apply makeshift screens made from found materials such as sheet metal, tank tracks and wire mesh. As shaped-charge weapons needed to impact armour at a set distance to impart maximum destruction, these 'bedsprings' acted to prematurely dissipate the narrow 500°C jet before it contacted the vehicle's main armour. Without them the crew risked being engulfed in a molten inferno that externally left a scorched hole dubbed a 'witch's kiss'.

As the replacement staff for Artillery Regiment zV constricted to new positions at Hohenwalde, Karlsaue, Karlsburg, Wardin and Helpe, the Soviet 88th Guards Heavy Tank Regiment (five IS-2s), 85th Individual Tank Regiment (eight T-34/85s), 43 open-topped SU-76 self-propelled guns and motorised artillery began moving into the area. Just behind, IX Guards Rifle Corps and XVIII Rifle Corps pushed north, but XII Guards Tank Corps was running out of fuel.

Tiger IIs knocked out several enemy tanks around Schönwerder, while more such attacks occurred south of Arnswalde. Intent on their own survival, many Nazi Party officials and the police abandoned Arnswalde for Reetz. Voigt, understandably incensed by the abandonment of the civilians, did what he could to get them out of the enveloping perimeter before it was too late. By the end of the day Tiger IIs had repulsed enemy probes near Neuwedell, where 503rd Heavy SS-Panzer Battalion's Workshop Company was located to be close to the fighting.

The 3rd of February brought warmer temperatures and a thaw that softened ground and made movement increasingly difficult, especially for motorised formations. South-west of Arnswalde, Tiger IIs rescued some surrounded infantry at Kopplinsthal, and four vehicles from 1st Panzer Company continued on to bolster Arnswalde's defences at Hohenwalde. Three Tiger IIs were damaged by heavy enemy armour and anti-tank fire. The remaining four vehicles fought near a wooded area at Sammenthin, where at 0700hrs the following morning Tiger 111 was destroyed and SS-Under Storm Leader Karl Brommann's vehicle (221) was immobilised by anti-tank fire and towed by three Tiger IIs to St Mary's Church in the centre of Arnswalde. Armoured train No. 77 from Group Munzel provided intermittent local support, but when the Soviets overran the tracks east of Reetz later on 4 February, it was forced from the area.

ARNSWALDE ENCIRCLED

With Soviet forces having reached the Ihna River south-west of Zachan, Steiner ordered Group Munzel to strengthen Arnswalde's defences. On 6 February the 11th SS-Volunteer Panzergrenadier Division Nordland sent in 15 StuG assault guns from 11th SS-Assault Gun Battalion and SS-Senior Storm Command Leader Paul-Albert Kausch's 11th SS-Panzer Battalion Hermann von Salza, which held off advancing enemy forces around Reetz. Soviet pressure on either flank, however, proved irresistible as masses of refugees tried to extricate themselves. When the Arnswalde–Reetz road was severed later in the day, upwards of 5,000 refugees were trapped within the Arnswalde perimeter. Heavy Soviet artillery made the situation seem all the more hopeless, and Voigt considered capitulation. In preparation for a rescue of the Arnswalde garrison, SS-Under Storm Leader Fritz Kauerauf (commander of 2nd Platoon, 1st SS-Panzer Company) was ordered to take three repaired Tiger IIs from the battalion workshop now at Stargard.

IS-2 SPECIFICATIONS

Production run: April 1944–June 1945 (15 months)

Vehicles produced: 4,392 (plus 107 IS-1s)

Combat weight: 46.08 tonnes (53 per cent armour weight)

Crew: four (commander, gunner, loader, driver)

DIMENSIONS

Length (hull/overall): 6.77m/9.83m

Width: 3.07m

Height: 2.73m

Ground clearance: 470mm

ARMOUR (THICKNESS AT DEGREES FROM VERTICAL)

Glacis (upper/lower): 120mm at 60°/120mm at 30°

Hull side (upper/lower): 90mm at 15°/90mm at 0°

Hull rear (upper/lower): 60mm at 49°/60mm at 41°

Hull roof: 30mm at 90°

Hull bottom: 20mm at 90°

Turret mantlet: 100mm (round)

Turret side: 90mm at 18°

Turret rear: 90mm at 30°

Turret roof: 30mm at 85–90°

Cupola (side/top): 90mm at 0°/20mm at 90°

ARMAMENT

Main gun: 122mm (121.92mm) Model 1943 D-25T L/43; 28 rounds – typically 20 OF-471/OF-471N (HE) and 8 BR-471 (APHE)

Sight: TSh-17 articulated telescope (4×); Mk. IV periscope

Secondary: 3 × 7.62mm DT machine guns (2,331 rounds for hull, coaxial and turret rear)

Main gun rate of fire: 2–3rpm

COMMUNICATIONS

Internal: TPU-4-BisF intercom

External: 10-R; later 10-RK (W/T and R/T stationary ranges of 24km and 16km, respectively)

MOTIVE POWER

Engine: 12-cylinder (water-cooled) 38.9 litre (diesel) (V-2-IS or V-2-K)

Power-to-weight: V-2-IS: 600hp at 2,300rpm (13hp/tonne) or V-2-K: 520hp at 2,200rpm (11.3hp/tonne)

Transmission: Synchromesh (clutch synchroniser); eight forward, two reverse gears

Fuel capacity: 790 litres (520 litres plus 3 × 90-litre external tanks)

PERFORMANCE

Ground pressure: 0.81kg/cm^2

Maximum speed (road/cross-country): 37kph/19kph

Operational range (road/cross-country): 150km (230km with external tanks)/120km (185km with external tanks)

Fuel consumption (road/cross-country): 3.5 litres/km / 4.3 litres/km

The next day Soviet forces pushed back the Dutch brigade (soon rechristened 23rd SS-Volunteer Panzergrenadier Division Nederland), overran Reetz and Hassendorf and cut road 104 to Stettin. Considerable Soviet forces, including armour and artillery, snaked their way north along the ridgeline just east of the Ihna River.

At dawn on 8 February Kausch ordered Kauerauf to send one of his Tiger IIs from 1st Platoon, 3rd SS-Panzer Company, and three StuGs under the experienced SS-Senior Storm Leader Hermann Wild to report on Soviet activity north of Reetz. After moving from Kausch's headquarters south of Jakobshagen, the group crested the high ground just west of Ziegenhagen and saw a seemingly endless enemy column of armour, artillery and infantry passing through Klein Silber, which if left unchecked threatened to move on to the Baltic coast and cut German forces still moving by land into Pomerania from the east.

OPPOSITE Pomeranian Wall: a Tiger II in the marketplace of Arnswalde (Neumark) in front of St Mary's Church, mid-February 1945. (Ullstein Picture via Getty Images)

Wild went for reinforcements, and soon returned with two Tiger IIs, ten additional StuGs from Battalion Herman von Salza, and a *Fallschirmjäger* company as a force with which to eliminate the enemy thrust.

Organising itself quickly, the German battle group halted to fire on several Soviet anti-tank guns west of Ziegenhagen around noon. As the *Fallschirmjäger* moved up on either side of the road past Ziegenhagen and across a bridge into Klein Silber, a pair of StuGs led a Tiger II through the latter town amid heavy small-arms fire. The assault guns were soon stopped by a Soviet anti-tank gun near a church some 200m distant, but a ridge in no man's land obscured each side and forced their fire high.

Kauerauf was apprised of the impasse and moved his taller Tiger II into a hull-down position to knock out the enemy gun crew with a Sprgr 43 HE round. As the German armour began to move on, Kauerauf was soon halted by a hastily laid enemy minefield in the street. The *Fallschirmjäger* fought their way forward, but with no engineers available one of the paratroopers rushed forward to destroy the mines with grenades and demolition charges. No sooner was the street cleared than a Soviet IS-2 came into view at 50m, which Kauerauf disabled with a Pzgr 39/43 anti-tank projectile, finishing it off with two more such hits. Two more IS-2s ground to a halt nearby after seeing Kauerauf's results and simply abandoned their vehicles and disappeared. With the Red Army's northward advance through Klein Silber now severed, the trio of Tiger IIs formed a defensive hedgehog formation along the village's southern end to take on fuel and ammunition. Over the next day two were destroyed by Soviet infantry and a third by its crew after it became immobilised.

As the front line began to stabilise along the Ihna River's southern edge, Soviet forces focused on eliminating the Arnswalde garrison. At 1000hrs on 9 February, eight Tiger IIs led ten armoured personnel carriers from 1st Battalion, 100th Panzergrenadier Regiment (Führer Escort Division), from the bridgehead at Fahrzoll, but the effort to reach the town's defenders stalled. With fuel and aircraft running low, Air Fleet (*Luftflotte*) 6 had several of its venerable Ju-52 transports airdrop supplies to the beleaguered 'fortress' during the nights of 8/9 February to 11/12, 13/14 and 14/15. Whether because of oversight or sabotage by foreign factory workers, much of the ammunition replenishment for the Tiger IIs comprised 8.8cm Flak 36 rounds, which being designed for anti-aircraft guns were unusable.

After the Red Army captured perhaps the most formidable section of the Pomeranian Wall at Deutsch Krone on the 11th, and Tiger IIs with Battalion Gross devastated a T-34 unit at Kähnsfelde, fighting around Arnswalde waned as both sides regrouped. To test the garrison's resolve and avoid a costly fight, the Soviets sent three German captives to Arnswalde's eastern perimeter at Springwerder during the evening of the 12th. Under a flag of truce the trio carried a message from their captors stating that to prevent unnecessary casualties the garrison needed to surrender by 0800hrs the following day. As an incentive the German defenders were told they would then be given food and medical attention, and be allowed to retain their personal effects, and civilians would simply be allowed to go their own way. But knowing the harsh fate that Germans in areas overrun by the Red Army had already experienced, there was little reason for them to expect different treatment.

At the designated time, instead of a white flag, the German defenders defiantly displayed both the German imperial and Nazi Party flags from St Mary's Church. Incensed, the Soviets unleashed an artillery, mortar and

OPPOSITE A Tiger II defends the Arnswalde perimeter near Kähnsfelde. On February 10, 1945, Tiger IIs, with support from Escort Battalion zbV Reichsführer-SS, stopped a Soviet assault on the Arnswalde perimeter at Kähnsfelde. To make the best use of the Tiger II's long range, these vehicles were positioned along the small hills that bordered the area's low-lying, swampy terrain and its small bisecting stream, the Stübenitz. With ammunition running low the German commander had to be very selective in choosing targets. Because of their firepower and protection, IS-2s would be a primary focus, but the more numerous medium T-34/85s could not be ignored. (Peter Dennis, © Osprey Publishing)

TIGER II GUNSIGHT

A Tiger II gunner uses his TzF 9d sight (through which this view is taken) to target an IS-2 at a range of 1,800m. Fine-tuning his aim from 2.5× to 5× magnification, the gunner prepares to fire a Pzgr 39/43 APCBC/HE-T round into the unsuspecting enemy vehicle's thinner side armour.

Believing themselves to be safely out of enemy range, the two halted IS-2s are taking on supplies and fuel before moving up for an attack. Soviet support personnel go about their business next to an American IHC M-5-6×4-318 supply truck.

If the target is beyond some 500m, the Tiger II's gunner estimates the target's actual size and divides it by the number of mills it encompasses in the scope. The loader adjusts the tick marks on the TzD 9d monocular gunsight in accordance with both the selected ammunition and the range, the latter being agreed upon by the driver, gunner and commander. Once it is rotated so that the large black triangle at the scope's top points to the estimated range, the upper tip of the large central triangle atop the vertical line is located between the targeted IS-2's turret and hull.

Katyusha rocket barrage lasting over seven hours that did considerable damage to the town. To the east, Schneidemühl's garrison faced imminent destruction. Organising into three groups, the garrison broke out for friendly lines on the 13th, but a quick Soviet response meant only a few reached their goal at Deutsch Krone. The roughly 15,000 civilians were left to the mercy of the Soviet and Polish forces that would take the city the next day.

Having gathered the 16 operational Tiger IIs that were not encircled, Eleventh SS-Panzer Army commanders were determined to prevent such a situation at Arnswalde. With most of III (Germanic) SS-Panzer Corps having been successfully transported by sea from the Courland Peninsula, Steiner was able to position sizeable forces east of Stargard. To lead Eleventh SS-Panzer Army's scheduled counter-attack, personnel of 11th SS-Volunteer Panzergrenadier Division Nordland conducted training and familiarised themselves with the area in the days leading up to the offensive before being put on alert on 14 February.

SONNENWENDE

With relatively warm temperatures continuing, intermittent rain and sleet greeted Eleventh SS-Panzer Army as it prepared to go over to the offensive between Lake Madü and Hassendorf. Although Steiner favoured first building his command's offensive capability, Guderian overruled him, and the attack was scheduled to start on the 16th.

In the grey pre-dawn Division Nordland's depleted 24th SS-Panzergrenadier Regiment Danmark moved its 2nd Battalion up to its jumping-off positions just south of the Ihna River. Regiment Danmark's Danish volunteers began their attack at 0600hrs with the intent of relieving Arnswalde's garrison as a prelude to the wider effort to sever the spearheads of Second Guards Tank Army.

By day's end companies from the 27th SS-Volunteer Division Langemarck had established positions before Petznick and Regiment Danmark held positions near Bonin where their patrols made contact with Voigt's encircled command.

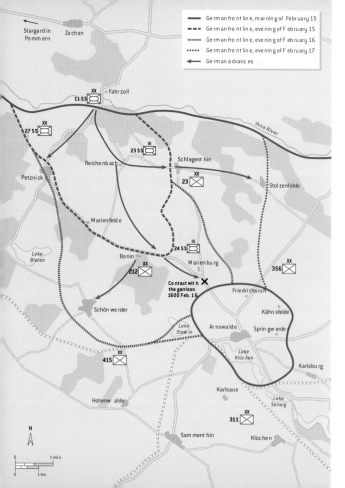

On Friday 16, Operation *Sonnenwende* officially commenced. On the German right XXXIX Panzer Corps' Panzer Division Holstein and the nearly full-strength 10th SS-Panzer Division Frundsberg pushed the 34th Guards Mechanised and 48th Guards Tank brigades back to south of Lake Madü. The Division Frundsberg's subsequent efforts to link up with 4th SS-Police Panzergrenadier Division, which with 28th SS-Volunteer Grenadier Division Wallonien was to form a pocket around Soviet forces between Arnswalde and the lake, came to naught. A quick and stubborn reaction by 66th Guards Tank Brigade with some 15 T-34/85s made further German progress near the area's saltworks difficult.

The southern Belgians from Division Wallonien made every effort to support Division Nordland's right flank, but could make little headway past Lake Plöne. Along Eleventh SS-Panzer Army's left X SS-Corps began its own offensive around Reetz, initially making good progress even though confronted by considerable Soviet anti-tank defences.

Constituting the primary German effort in the centre, III (Germanic) SS-Panzer Corps crossed the Ihna River to reinforce the Division Nordland's success the previous day. The Soviet VII Guards Cavalry Corps fell back in some disarray, but XVIII Rifle Corps continued to maintain its grip around Arnswalde. Until greater numbers of heavy Soviet artillery could be redirected northwards, the front line was not likely to solidify any time soon. Sixty-First Army commander Colonel-General Pavel Belov moved to reduce the Arnswalde garrison before the Germans could break the siege by sending two heavy tank regiments from 11th Guards Heavy Tank Brigade to the area as a breakthrough element for the 356th and 212th Rifle divisions. Having only 260 and 300 soldiers available for action, respectively, the force was in poor shape to undertake such a mission. The 85th Tank and 1899th Self-Propelled Artillery regiments were subsequently moved up to provide armoured support for a reimplemented attack that included the 311th and 415th Rifle divisions once they arrived on the scene.

Tiger IIs from 503rd Heavy SS-Panzer Battalion were able to effectively engage enemy armour at long range, but the recent thaw had created very muddy terrain that hindered movement. The Germans knew that if the ground supported a man standing on one leg and carrying another man on his shoulders, it would support a tank. Although considerably reduced in strength by recent action along the Baltic coast, Division Nordland continued to fight through sporadic Soviet resistance. Retaining the element of surprise, German forces exploited the situation to resolve their mission as quickly as possible in spite of environmental conditions unsuited for armoured operations. In response to the broader thrust of Operation *Solstice*, Soviet commanders activated several local IS-2 formations for a counter-attack. Lieutenant-Colonel Joseph Rafailovich's 70th Guards Heavy Tank Regiment (Forty-Seventh Army) was positioned north of Woldenberg, while Lieutenant-Colonel Semen Kalabukhov's 79th (XII Guards Tank Corps) assembled near Dölitz. Lieutenant-Colonel Peter Grigorevich's

88th Guards Heavy Tank Regiment was also available near Berlinchen in Sixty-First Army's sector.

The 24th SS-Panzergrenadier Regiment Danmark's 3rd Battalion was ordered to take Bonin with support from Führer Escort Division's mounted Panzergrenadiers and three StuGs from Division Nordland. On establishing a defensive position south of the village and the nearby Volkswagen factory, the half-tracks and assault guns turned for Schönwerder to assist Regiment Danmark's 2nd Battalion. The 1st Battalion, 66th SS-Grenadier Regiment (Division Langemarck) launched a concurrent attack with Regiment Danmark's 3rd Battalion's effort to take Marienfelde and establish a combat outpost at Petznick.

At 1600hrs parts of Regiment Danmark broke from their positions at Bonin and captured Schönwerder in a quick assault. As additional companies reinforced the success, the remainder of the battalion weathered XVIII Rifle Corps' artillery fire to reach Arnswalde's north-western perimeter and break the 11-day siege with a defensible corridor. Seven Tiger IIs that had been attached to Division Nordland soon entered the town along with other reinforcements that proceeded to strengthen the garrison's defences. As the Germans expanded the corridor, strong enemy resistance along the Stargard–Arnswalde railway stopped further progress in that area. To the north contact was made with parts of Regiment Danmark's 3rd Battalion along the tracks with friendly units at Marienburg.

On the 17th, Second Guards Tank Army arrived in force in the Arnswalde sector and stopped what little impetus the Frundsberg and SS-Police divisions retained. While the latter tried to work into the flank and rear of XII Guards Rifle Corps and 75th Rifle Division, 6th Guards Heavy Tank Regiment moved up to counter the effort.

The Division Wallonien continued to hold on to their positions in the Linden Hills, with one company fighting nearly to the last man against repeated enemy assaults. Around Arnswalde, 14 IS-2s from Hero of the Soviet Union Major Prokofi Kalashnikov's 90th Guards Heavy Tank Regiment moved up to join the

TIGER II, 505TH HEAVY PANZER BATTALION, THÜRINGEN, 1944

After heavy fighting during Operation *Bagration*, the 505th was ordered off the Eastern Front to rest and refit at the troops training grounds at Ohrdruf (Thüringen) in July and late August 1944. Here it received its first Tiger IIs. The unit was responsible for some of the most spectacular non-regulation markings that were applied to Tiger IIs: they removed a rectangle of *Zimmerit* from the turret side and therein painted their unit emblem, the knight on a charger.

The exact colours of the emblem cannot be confirmed at this time as there may have been different colours for each company. The call sign, normally on the turret side, was painted on the gun mantle and barrel. The call sign '213' was repeated on the turret rear escape hatch. Tiger IIs of the *Stabskompanie* used the roman numerals I, II, and III similarly located.

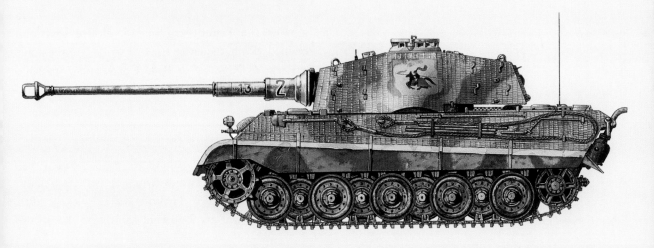

already engaged 91st and 92nd with six and five IS-2s respectively, but success remained elusive. The 356th Rifle Division managed to get infantry elements into the suburbs near the city's gas works, but German infantry firing from the upper floors of buildings, and roaming Tiger IIs, made infiltration impossible.

With a stable corridor out of Arnswalde now available, Voigt quickly orchestrated the evacuation of the civilians and wounded. Although it had been briefly severed, Soviet efforts to permanently eliminate the 2km-wide corridor were unsuccessful. Of the original seven garrison Tiger IIs, only four remained operational. Battered and in need of maintenance, the vehicles were withdrawn to Zachan.

AFTERMATH AND ANALYSIS

While medium armoured fighting vehicles such as the German Panther or Soviet T-34 possessed a balanced triad of firepower, mobility and protection that permitted them to undertake a variety of combat roles, the Tiger II's greater weight relegated it to more limited defensive operations. Its size made movement through urban environments or along narrow roads difficult. Moreover, its drivetrain was under-strength, the double radius L801 steering gear was stressed and the seals and gaskets were prone to leaks. Limited crew training could amplify these problems as inexperienced drivers could inadvertently run the engine at a high rpm or move over terrain that overly taxed the suspension. Extended travel times under the Tiger II's own power stressed the swing arms that supported the road wheels and made them susceptible to bending. Such axial displacement would probably strain the tracks and bend the link bolts to further disrupt proper movement. Overworked engines needed to be replaced roughly every 1,000km.

Although wide tracks aided movement over most terrain, should the vehicle require recovery another Tiger II was typically needed to extract it. Requirements for spare parts were understandably high, and maintenance was an ongoing task, all of which reduced vehicle availability.

The Tiger II's long main armament, the epitome of the family of 8.8cm anti-aircraft/anti-tank guns that had terrorised enemy armour since the Spanish Civil War of 1936–39, fired high-velocity rounds along a relatively flat trajectory. In combination with an excellent gunsight, the weapon system was accurate at long range, which enabled rapid targeting and a high first look/first hit/first kill probability. However, the lengthy barrel's overhang stressed the turret ring, and made traverse difficult when not on level ground. Optimally initiating combat at distances beyond which an enemy's main armament could effectively respond, the Tiger II's lethality was further enhanced by its considerable armour protection, especially across the frontal arc that provided for a high degree of combat survivability. Although the vehicle's glacis does not appear to have ever been penetrated during battle, its flanks and rear were vulnerable to enemy anti-tank weapons at normal ranges.

In the hands of an experienced crew, and under environmental and terrain conditions that promoted long-range combat, the weapon system achieved a high kill ratio against its Allied and Red Army counterparts. The 503rd Heavy SS-Panzer Battalion, for example, was estimated to have scored an estimated 500 'kills' during the unit's operational life from January to April 1945. While such a figure was certainly inflated by inaccurate record keeping, it illustrates the success of the weapon system if properly employed and supported. Of 503rd Heavy SS-Panzer Battalion's original complement of 39 Tiger IIs, only ten were destroyed through combat, with the remainder being abandoned or destroyed by their crews due to mechanical breakdowns or lack of fuel. As the unit never received replacement tanks (in contrast to its sister 501st and 502nd battalions, which were given 2.38 and 1.7 times their respective vehicle allotments), its Tiger II combat losses averaged less than 50 per cent.

Because of the chaotic combat environment throughout Pomerania, and the need to quickly allocate resources to several threatened sectors at once, the Tiger IIs were frequently employed singly, or in small groups, often at the will of a local senior commander. In much the same way as with the French in 1940, 503rd Heavy SS-Panzer Battalion's armour acted more in an infantry-support capacity than as a unified armoured fist. The Tiger IIs would perhaps have been better used organisationally to fill a Panzer regiment's heavy company by strengthening existing, depleted parent formations; but instead they remained in semi-independent heavy Panzer battalions until the end of the war. Forced to rely on small-unit tactics, Tiger II crews played to their strengths by adopting ambush tactics to minimise vehicular movement and pre-combat detection, especially from enemy ground-attack aircraft.

As tankers regularly spent long hours in their mounts, the Tiger II's relatively spacious interior helped reduce fatigue, and made operating and fighting within the vehicle somewhat less taxing. A good heating and ventilation system improved operating conditions, which then reduced crew mistakes that were all too common during a chaotic firefight. Although the Tiger II had well-positioned ammunition racks that facilitated loading, projectiles that were stored in the turret bustle were susceptible to potentially catastrophic damage caused by spalling or projectile impacts. Even after Henschel incorporated spall liners to reduce such debris, concerned crews would often leave the turret rear empty, which correspondingly made room to use the rear hatch as an emergency exit.

The cost to produce the Tiger II in manpower and time (double that of a 45-tonne Panther), and its high fuel consumption, brought into question why such a design progressed beyond the drawing board, considering Germany's dwindling resources and military fortunes. It was partly a response to the perpetual escalation of the requirement to achieve or maintain battlefield supremacy, and much of the blame rested with Hitler and his desire for large armoured vehicles that in his view presumably reflected Germany's might and reinforced

OPPOSITE Eastern Front, December 30, 1944: camouflaged Tiger II tanks lie in wait in a ravine. (Bundesarchiv)

propaganda. By not focusing resources on creating greater numbers of the latest proven designs such as the Panther G, German authorities showed a lack of unified direction and squandered an ability to fight a war of attrition until it was too late to significantly affect the outcome. Limited numbers of qualitatively superior Tiger IIs could simply not stem the flood of enemy.

THE IS-2 COMPARED

When the Red Army transitioned to the strategic offensive in early 1943, their skill in operational deception increasingly enabled them to mass against specific battlefield sectors, often without the Germans realising the degree to which they were outnumbered until it was too late. Purpose-built to help create a breach in the enemy's front-line defences, the IS-2's relatively light weight, thick armour and powerful main gun made the design ideal for such hazardous tip-of-the-spear operations. Once a gap was created, follow-up armoured and mechanised formations were freed to initiate the exploitation and pursuit phase of the Red Army's 'deep battle' doctrine. Here, more nimble vehicles such as the T-34 could concentrate on moving into an enemy's flank and rear areas to attack their logistics and command and control capabilities. Unlike the mechanically unreliable KV-1, the IS-2 had a surprisingly high life expectancy of some 1,100km. Able to cover considerable distances on its own, as evidenced during the Vistula–Oder offensive, it remained at the fore of Soviet offensive operations until the end of the war.

With some 20 HE rounds out of an ammunition complement of 28, the IS-2 was well suited to attacking targets such as fortifications, buildings, personnel and transport vehicles. In this capacity its heavy, two-piece ammunition and slow loading and reaction times were not much of an issue. Against armour such delays could prove catastrophic. Should the 122mm D-25T gun score a hit, however, its relatively low-velocity projectiles imparted considerable force that could severely damage what they could not penetrate. Having a large quantity of low-grade propellant, its rounds created smoke that could reveal its firing position. Its periscope did not provide all-around viewing or quick targeting reconciliation, and traversing the large, overhanging barrel was often impeded when the vehicle was not on level terrain. Thoughts of increasing the mantlet thickness were stillborn as any additional weight to the turret's front would only exacerbate the problem. In a potentially fast-paced tank battle, where being first to get a round on the target generally decided the contest, the IS-2 was often at a disadvantage.

As the IS-2's glacis presented a fairly small surface area, its turret was the most likely target for enemy anti-tank weapons. To defend against German handheld weapons such as the Panzerfaust and Panzerschreck, IS-2 crews made increasing use of ad hoc metal turret screens or skirts to disrupt the effect of the warhead's shaped-charge. Although prone to deformation in cramped urban environments, such protection was better than nothing.

The IS-2's rugged construction was more forgiving than that of its Tiger II rival. Crew safety and comfort were secondary considerations to Soviet tank designers throughout the war, as evidenced by the rough application of welding and seaming, and construction techniques. Like the mass-produced T-34, the IS-2 was a relatively simple, rugged design that facilitated construction and in turn provided numbers to overwhelm the enemy and help bridge the gap between Soviet and German capabilities after a long period of German technological superiority. In the fighting around Arnswalde IS-2s possessed better mobility, but the relatively open terrain and limited foliage plus the greater range of the Tiger II largely prohibited the Soviet tanks from getting close to the town – and the Germans focused on enemy tanks over other targets.

CONCLUSION

Despite the many drawbacks of expense, mechanical difficulties and problems of mobility and transportation, the Tiger I and II undoubtedly deserved their legendary status acquired on the battlefields of World War II.

The combination of powerful guns and armour so heavy as to be impenetrable to all but a few opponents was, as Hitler intended, a fearsome one, but ultimately the low production rate, especially of the heavy Tiger II, and the vulnerability of both tanks to mechanical problems requiring intensive maintenance, recovery and repair, meant that the Tiger I and II's effectiveness as war-winning weapons was drastically limited.

As the war progressed, the difficulty in developing, fielding and supporting a variety of armoured vehicle sizes and types became readily apparent to nations, who found the resulting mix of light, medium and heavy tanks and self-propelled guns (often with overlapping and redundant capabilities and incompatible parts) cost-prohibitive to maintain. The single multirole main battle tank design, capable of being mass-produced and able to be modified and upgraded as needed, became increasingly appealing, and the race to build the most indestructible heavy tank was no longer a plausible expenditure of effort.

Although soon outmoded, the Tiger I and II at their best were as formidable as the myths around them suggested. That they still continue to capture the imagination today is attested to by their enduring popularity among armour enthusiasts, modellers and gamers, and the star power of those few examples maintained in running order at museums around the world. Chief among these are Tiger 131 of the 504th Heavy Panzer Battalion, on show at the Tank Museum, Bovington, UK, and the only Tiger II in running order, on display at the Musée des Blindés, Saumur, France.

OPPOSITE The only running Tiger I tank in the world, at Tankfest 2013 at the Tank Museum at Bovington. (Getty)

GLOSSARY AND ABBREVIATIONS

AFV armoured fighting vehicle

AG *Aktiengesellschaft*, ('corporation')

Angriff nach Vorbereitung deliberate attack

AP armour piercing

APCBC armour-piercing capped ballistic capped

APCR armour-piercing composite rigid

APDS armour-piercing discarding sabot

Ausf. *Ausführung* ('design')

Befehlswagen command tank

BHN Brinell Hardness Number

Breitkeil broad-wedge formation

DHHV Dortmund Hörder Hutten Verein

Doppelreihe double row

DW *Durchbruchwagen*

Elfenbein ivory

Fallschirmjäger paratroopers

FDL forward defensive locality

Feldgrau 'field grey', a grey-green

Fgst *Fahrgestell* (chassis)

Fkl *Funklenk* (radio controlled)

FlaK *Flugzeugabwehrkanone* (aircraft defence cannon)

FuG *Funk Gerät* (radio set)

Gr *Granate* (shell)

HE high explosive

HEAT high-explosive anti-tank

Heeres-Dienstvorschrift Army Service Regulation

Heereswerkmeister army specialists

Heereszeugamt Army Ordnance Department

HE-T high explosive-tracer

HL *Hochleistungsmotor*

HVAP-T hyper-velocity armour piercing-tracer

IS Iosef Stalin (tank)

Keil wedge

KV Kliment Voroshilov (tank)

KwK *Kampfwagenkanone* ('fighting vehicle cannon')

Landesschützen Territorial Defence

Linie line

maskiróvka deceptive measures

MG machine gun

Nahverteidigungswaffe Close Defence Weapon

NCO non-commissioned officer

Panzerwaffe Armoured Arm

Panzerwarte tank mechanics

Pilz ('mushroom') sockets

Prismenspiegelkuppel commander's cupola

Pzgr *Panzergranate* (armour-piercing shell)

RAF Royal Air Force

Reihe row

RHA rolled homogeneous armour

ROQF Royal Ordnance Quick-Firing

Saukopf 'pig's head' (mantlet)

SGP Simmering-Graz-Pauker

Sofortangriffe quick attacks

SPG self-propelled gun

Sprgr *Sprenggranate* (HE round)

Taktzeiten cycle times – *Takt* for short

Tigerfibel Tiger Manual (Field Service Regulation D656/27)

TzF *Turmzielfernrohr*

VK *Versuchsfahrgestelle* (test chassis)

Volkssturm 'People's storm': late-war German militia

Vorbut meeting engagement

Waffenamt WaA – the Army Weapons Agency

Waffenprüfamt Weapons Proving Office

zbV *zur besonderen Verwendung* ('for special employment')

Zimmerit anti-magnetic paste

zV *zur Vergeltung* ('for retribution')

INDEX